Engaging the **Private-Sector Health Care System** in Building Capacity to Respond to Threats to the **Public's Health and National Security**

PROCEEDINGS OF A WORKSHOP

Joe Alper, *Rapporteur*

Forum on Medical and Public Health Preparedness for Disasters and Emergencies

Board on Health Sciences Policy

Health and Medicine Division

The National Academies of
SCIENCES · ENGINEERING · MEDICINE

THE NATIONAL ACADEMIES PRESS
Washington, DC
www.nap.edu

THE NATIONAL ACADEMIES PRESS   500 Fifth Street, NW   Washington, DC 20001

This activity was supported by a contract between the National Academy of Sciences and the Department of Health and Human Services (HHSO100201750005A). Any opinions, findings, conclusions, or recommendations expressed in this publication do not necessarily reflect the views of any organization or agency that provided support for the project.

International Standard Book Number-13:   978-0-309-48212-7
International Standard Book Number-10:   0-309-48212-7
Digital Object Identifier: https://doi.org/10.17226/25203

Additional copies of this publication are available for sale from the National Academies Press, 500 Fifth Street, NW, Keck 360, Washington, DC 20001; (800) 624-6242 or (202) 334-3313; http://www.nap.edu.

Copyright 2018 by the National Academy of Sciences. All rights reserved.

Printed in the United States of America

Suggested citation: National Academies of Sciences, Engineering, and Medicine. 2018. *Engaging the private-sector health care system in building capacity to respond to threats to the public's health and national security: Proceedings of a workshop*. Washington, DC: The National Academies Press. doi: https://doi.org/10.17226/25203.

The National Academies of
SCIENCES • ENGINEERING • MEDICINE

The **National Academy of Sciences** was established in 1863 by an Act of Congress, signed by President Lincoln, as a private, nongovernmental institution to advise the nation on issues related to science and technology. Members are elected by their peers for outstanding contributions to research. Dr. Marcia McNutt is president.

The **National Academy of Engineering** was established in 1964 under the charter of the National Academy of Sciences to bring the practices of engineering to advising the nation. Members are elected by their peers for extraordinary contributions to engineering. Dr. C. D. Mote, Jr., is president.

The **National Academy of Medicine** (formerly the Institute of Medicine) was established in 1970 under the charter of the National Academy of Sciences to advise the nation on medical and health issues. Members are elected by their peers for distinguished contributions to medicine and health. Dr. Victor J. Dzau is president.

The three Academies work together as the **National Academies of Sciences, Engineering, and Medicine** to provide independent, objective analysis and advice to the nation and conduct other activities to solve complex problems and inform public policy decisions. The National Academies also encourage education and research, recognize outstanding contributions to knowledge, and increase public understanding in matters of science, engineering, and medicine.

Learn more about the National Academies of Sciences, Engineering, and Medicine at **www.nationalacademies.org**.

The National Academies of
SCIENCES • ENGINEERING • MEDICINE

Consensus Study Reports published by the National Academies of Sciences, Engineering, and Medicine document the evidence-based consensus on the study's statement of task by an authoring committee of experts. Reports typically include findings, conclusions, and recommendations based on information gathered by the committee and the committee's deliberations. Each report has been subjected to a rigorous and independent peer-review process and it represents the position of the National Academies on the statement of task.

Proceedings published by the National Academies of Sciences, Engineering, and Medicine chronicle the presentations and discussions at a workshop, symposium, or other event convened by the National Academies. The statements and opinions contained in proceedings are those of the participants and are not endorsed by other participants, the planning committee, or the National Academies.

For information about other products and activities of the National Academies, please visit www.nationalacademies.org/about/whatwedo.

PLANNING COMMITTEE ON ENGAGING THE PRIVATE-SECTOR HEALTH CARE SYSTEM IN BUILDING CAPACITY TO RESPOND TO THREATS TO THE PUBLIC'S HEALTH AND NATIONAL SECURITY[1]

HELEN BURSTIN (*Co-Chair*), Executive Vice President and Chief Executive Officer, Council of Medical Specialty Societies
ARTHUR L. KELLERMANN (*Co-Chair*), Professor and Dean, F. Edward Hebert School of Medicine, Uniformed Services University of the Health Sciences
MICHAEL ANDERSON, President, UCSF Benioff Children's Hospital and Senior Vice President, Children's Services, UCSF Health, University of California, San Francisco
ANGELA BRICE-SMITH, Deputy Consortium Administrator for Quality Improvement and Survey & Certification Operations, Centers for Medicare & Medicaid Services
ERIC EPLEY, Executive Director, Southwest Texas Regional Advisory Council
PAUL HINCHEY, Assistant Medical Director, Wake County EMS
THOMAS KIRSCH, Director, National Center for Disaster Medicine and Public Health and Professor of Military and Emergency Medicine, F. Edward Hebert School of Medicine, Uniformed Services University of the Health Sciences
JON KROHMER, Director, Office of Emergency Medical Services, National Highway Traffic Safety Administration, Department of Transportation
NICOLETTE LOUISSAINT, Executive Director, Healthcare Ready
RICARDO MARTINEZ, Chief Medical Officer, Adeptus Health
CARTER MECHER, Senior Medical Advisor, Office of Public Health, Department of Veterans Affairs
JOHN OSBORN, Operations Manager and Assistant Professor, Health Care Systems Engineering, Department of Surgery, Mayo Clinic College of Medicine
TODD RASMUSSEN, Associate Dean for Research and Harris B. Schumaker, Jr. Professor of Surgery, F. Edward Hebert School of Medicine, Uniformed Services University of the Health Sciences
ROSLYNE SCHULMAN, Director for Policy Development, American Hospital Association

---

[1] The National Academies of Sciences, Engineering, and Medicine's planning committees are solely responsible for organizing the workshop, identifying topics, and choosing speakers. The responsibility for the published Proceedings of a Workshop rests with the workshop rapporteur and the institution.

SKIP SKIVINGTON, Vice President of Healthcare Continuity Management and Support Services, Kaiser Permanente
JEFFREY UPPERMAN, Director, Trauma Program and Director, Pediatric Disaster Resource and Training Center, Children's Hospital of Los Angeles
MICHAEL WARGO, Assistant Vice President, Enterprise Preparedness & Emergency Operations, Hospital Corporation of America

Health and Medicine Division Staff

SCOTT WOLLEK, Senior Program Officer
CLAIRE GIAMMARIA, Associate Program Officer
BEN KAHN, Research Associate
MARIA BABIRYE, Senior Program Assistant (*until August 2018*)
ANDREW M. POPE, Director, Board on Health Sciences Policy

Consultants

JOE ALPER, Writer, LSN Consulting
LAURA RUNNELS, LAR Consulting

FORUM ON MEDICAL AND PUBLIC HEALTH PREPAREDNESS FOR DISASTERS AND EMERGENCIES[1]

DAN HANFLING (*Co-Chair*), Johns Hopkins Center for Health Security
SUZET MCKINNEY (*Co-Chair*), Illinois Medical District
STACEY ARNESEN, National Library of Medicine, National Institutes of Health
ERIC BLANK, Association of Public Health Laboratories
MARY CASEY-LOCKYER, American Red Cross
BROOKE COURTNEY, Office of Counterterrorism and Emerging Threats, Food and Drug Administration
JOHN DREYZEHNER, Association of State and Territorial Health Officials
DAVID EISENMAN, University of California, Los Angeles
BRUCE EVANS, National Association of Emergency Medical Technicians
LARRY FLUTY, Office of Health Affairs, Department of Homeland Security
JOHN HICK, Hennepin County Medical Center
ROBERT KADLEC, Assistant Secretary for Preparedness and Response, Department of Health and Human Services
CLAUDIA KELLY, Seqirus
THOMAS KIRSCH, Uniformed Services University of the Health Sciences, Department of Defense
DREW LEWIS, Meridian Medical Technologies
NICOLETTE A. LOUISSAINT, Healthcare Ready
FREDA LYON, Emergency Nurses Association
CAROLYN MEIER, Administration for Children and Families, Department of Health and Human Services
AUBREY MILLER, National Institute of Environmental Health Sciences, National Institutes of Health
JOHN OSBORN, Mayo Clinic College of Medicine
TARA O'TOOLE, In-Q-Tel
ANDREW PAVIA, Infectious Diseases Society of America
ALONZO PLOUGH, Robert Wood Johnson Foundation
TERRY RAUCH, Defense Health Agency, Department of Defense
STEPHEN REDD, Office of Public Health Preparedness and Response, Centers for Disease Control and Prevention
SARA ROSZAK, National Association of Chain Drug Stores
ROSLYNE SCHULMAN, American Hospital Association

---

[1] The National Academies of Sciences, Engineering, and Medicine's forums and roundtables do not issue, review, or approve individual documents. The responsibility for the published Proceedings of a Workshop rests with the workshop rapporteur and the institution.

RICHARD SERINO, Harvard University School of Public Health
ALAN SINISCALCHI, Council of State and Territorial Epidemiologists
CRAIG VANDERWAGEN, East West Protection, LLC
JENNIFER WARD, Trauma Center Association of America
GAMUNU WIJETUNGE, National Highway Traffic Safety Administration, Department of Transportation
MATTHEW WYNIA, University of Colorado Denver

Health and Medicine Division Staff

SCOTT WOLLEK, Senior Program Officer
LISA BROWN, Senior Program Officer
CLAIRE GIAMMARIA, Associate Program Officer
BEN KAHN, Research Associate
MARIA BABIRYE, Senior Program Assistant (*until August 2018*)
ANDREW M. POPE, Director, Board on Health Sciences Policy

Reviewers

This Proceedings of a Workshop was reviewed in draft form by individuals chosen for their diverse perspectives and technical expertise. The purpose of this independent review is to provide candid and critical comments that will assist the National Academies of Sciences, Engineering, and Medicine in making each published proceedings as sound as possible and to ensure that it meets the institutional standards for quality, objectivity, evidence, and responsiveness to the charge. The review comments and draft manuscript remain confidential to protect the integrity of the process.

We thank the following individuals for their review of this proceedings:

MAHSHID ABIR, University of Michigan
MICHAEL J. CONSUELOS, The Hospital & Healthsystem Association of Pennsylvania
JORIE KLEIN, Parkman Memorial Hospital
JEFFREY UPPERMAN, Children's Hospital Los Angeles

Although the reviewers listed above provided many constructive comments and suggestions, they were not asked to endorse the content of the proceedings nor did they see the final draft before its release. The review of this proceedings was overseen by **SARA ROSENBAUM,** George Washington University. She was responsible for making certain that an independent examination of this proceedings was carried out in accordance with standards of the National Academies and that all review comments were carefully considered. Responsibility for the final content rests entirely with the rapporteur and the National Academies.

Contents

ACRONYMS AND ABBREVIATIONS		xiii
1	**INTRODUCTION AND OVERVIEW**	1
	Sponsor's Charge, 2	
	Organization of the Proceedings, 5	
2	**PERSPECTIVES ON THE NATION'S CAPACITY TO RESPOND TO THREATS TO HEALTH, SAFETY, AND SECURITY**	7
	Private Health System Perspectives, 7	
	Discussion, 13	
	Federal Perspectives, 17	
	Discussion, 22	
3	**LEARNING FROM EXPERIENCE**	25
	Case Studies in Cross-Sector Collaboration from Past Disruptions, 25	
	Discussion, 34	
4	**ASPR'S NEW VISION FOR A REGIONAL HEALTH RESPONSE SYSTEM**	43
5	**LOOKING TO THE FUTURE**	51
	Cultivating Best Practices, 53	
	Discussion, 62	

Leading Change Across the Field, 65
Discussion, 70
Leading Change at the Local Level, 71
Discussion, 75

6 **EXPLORING OPPORTUNITIES TO IMPROVE PRIVATE-SECTOR INVESTMENT IN CAPACITY BUILDING**   81
Examining Regulatory Barriers and Facilitators, 81
Discussion, 83
Examining Financial Barriers and Facilitators, 85
Discussion, 88

7 **FINAL THOUGHTS**   95
Discussion, 96

REFERENCES   99

APPENDIXES
A Workshop Agenda   101
B Statement of Task   109
C Speaker Biographies   111

Acronyms and Abbreviations

AMR	American Medical Response
ASPR	Office of the Assistant Secretary for Preparedness and Response
BARDA	Biomedical Advanced Research and Development Authority
CBRN	chemical, biological, radiological, and nuclear
CDC	Centers for Disease Control and Prevention
CMOC	Catastrophic Medical Operations Center
CMS	Centers for Medicare & Medicaid Services
DHS	Department of Homeland Security
DMAT	disaster medical assistance team
DoD	Department of Defense
EMS	emergency medical services
FAA	Federal Aviation Administration
FEMA	Federal Emergency Management Agency
GIS	geographical information system
HCA	Hospital Corporation of America
HHS	Department of Health and Human Services

HIPAA	Health Insurance Portability and Accountability Act
HPP	Hospital Preparedness Program
NDMS	National Disaster Medical System
NERC	North American Electric Reliability Corporation
NGO	nongovernmental organization
NORAD	North American Aerospace Defense Command
NTSB	National Transportation Safety Board
PAHO	Pan American Health Organization
REPLICA	Recognition of EMS Personnel Licensure Interstate CompAct
SETRAC	SouthEast Texas Regional Advisory Council
SNS	Strategic National Stockpile
STRAC	Southwest Texas Regional Advisory Council
USNORTHCOM	U.S. Northern Command
USUHS	Uniformed Services University of the Health Sciences
VA	Department of Veterans Affairs
WebEOC	Web Emergency Operations Center

1

Introduction and Overview[1]

Disasters tend to cross political, jurisdictional, functional, and geographic boundaries. As a result, disasters often require responses from multiple levels of government and multiple organizations in the public and private sectors. This means that public and private organizations that normally operate independently must work together to mount an effective disaster response (Auf der Heide and Scanlon, 2007). To identify and understand approaches to aligning health care system incentives with the American public's need for a health care system that is prepared to manage acutely ill and injured patients during a disaster, public health emergency, or other mass casualty event, the National Academies of Sciences, Engineering, and Medicine hosted a 2-day public workshop on March 20 and 21, 2018. Titled Engaging the Private-Sector Health Care System in Building Capacity to Respond to Threats to the Public's Health and National Security, this workshop had the following objectives as developed by an ad hoc planning committee:

---

[1] The planning committee's role was limited to planning the workshop, and the Proceedings of a Workshop was prepared by the workshop rapporteur and staff as a factual summary of what occurred at the workshop. Statements, recommendations, and opinions expressed are those of individual presenters and participants, and are not necessarily endorsed or verified by the National Academies of Sciences, Engineering, and Medicine, and they should not be construed as reflecting any group consensus.

- Explore the degree to which the public and private health care systems self-identify as key components of the U.S. disasters and national security infrastructure;
- Discuss interest among health care institutions in developing collaborations across public and private sectors with the aim of strengthening capacity to respond to disasters and public health emergencies;
- Consider possible key levers that would motivate private-sector investment in system capacity building for disaster and public health emergency response, including those levers that already exist, but are not currently used as incentives to expand this capacity (quality measurement, grant programs, alternative payment models, tax benefits, etc.);
- Explore possible strategies to overcome key challenges to applying existing incentives to improve the quality, effectiveness, and efficiency of the management of critically ill and injured patients on a day-to-day basis and during emergency response scenarios;
- Review possible key sources of information and data elements that could be used to improve situational awareness of public- and private-sector health care facility capacity and capabilities to respond to disasters and public health emergencies; and
- Understand the degree to which Department of Defense (DoD) or Department of Veterans Affairs (VA) hospitals could be used as a part of the U.S. response to disasters and public health emergencies requiring a health care response.

The workshop agenda can be found in Appendix A, and the workshop's Statement of Task is detailed in Appendix B.

SPONSOR'S CHARGE

To begin the workshop, Kevin Yeskey from the Office of the Assistant Secretary for Preparedness and Response (ASPR) of the Department of Health and Human Services (HHS) provided some background on ASPR.[2] ASPR's mission, he said, is to save lives and protect Americans from 21st-century threats, many of which have the possibility of causing unimaginable health consequences. ASPR's charge includes planning for and responding to seasonal threats such as hurricanes, tornadoes, and debilitating snowstorms; infrequent threats, such as earthquakes; and new threats, including

---

[2] For the purposes of this publication, unless otherwise stated, ASPR refers to the Office of the Assistant Secretary for Preparedness and Response rather than the Assistant Secretary himself.

state-sponsored terrorism, mass shootings, and bombings. The new threats, noted Yeskey, were not a big part of the public consciousness when ASPR was established in 2007; they cause injuries that most hospitals and trauma centers do not see at a frequency that enables them to handle such events with optimal efficiency. "We need to not only act stronger and with more capacity, we need to act smarter," said Yeskey. "We need to build smartness and efficiencies into the way we respond and recover as well as prepare."

The question ASPR gets regularly is, "Are we ready?" The 2017 hurricane season, with three significant hurricanes in a 4-week period, tested the nation's emergency response and health care system capacity. Yeskey noted that HHS can call on the National Disaster Medical System (NDMS) at times of great need, but the volunteer force of federal employees that is part of NDMS was stretched to its limit by the time Maria struck Puerto Rico and the U.S. Virgin Islands. What Yeskey found alarming was that these were relatively straightforward events for which there were warnings and time to prepare, but some 21st-century threats do not have that fidelity as far as when and where they are going to strike. "We are going to have to look at how we can improve that system and improve our response capabilities," said Yeskey.

Beyond that, the events in the fall of 2017 illustrated and amplified the health care system's dependence on its supporting infrastructure. Hospitals, for example, depend on electricity, but the supplemental generators meant to keep hospitals with power during an outage are not built to run for days on end. In addition, as Yeskey noted, generators require fuel, and if roads are impassable, fuel may also be in short supply. "Those dependencies were very much thought about and discussed during the hurricanes, and we need a way to work through understanding those dependencies, whether it is in the supply chain or the electrical grid, and how to better get our health care facilities and our clinics back online and staffed and equipped and supplied," he said.

Moreover, emergencies of that scale need the entire community—businesses, faith-based organizations, nongovernmental organizations (NGOs), and local government—to be involved, which calls for all sectors to be engaged in planning, response, and recovery activities. Speaking from ASPR's perspective, Yeskey said, "We do not know people [in the community] and have not engaged in trusted relationships that will help us work through the complex problems associated with response and recovery." That realization was one of the main reasons that ASPR asked the Forum on Medical and Public Health Preparedness for Disasters and Emergencies to hold this workshop. "Our emphasis today and tomorrow is to talk about public–private partnerships, how to develop those and not engage in those transactional onesies and twosies kind of relationships, but have ongoing, credible, trusting relationships where you know the people you are dealing with, and can plan together and work together," said Yeskey.

As an example of the work ASPR is doing to develop better relationships in communities, Yeskey described the Critical Infrastructure Protection Program. This initiative works with private industry to address problems and identify solutions from the private sector for infrastructure protection. He also mentioned a coordinating council, consisting of representatives from the private sector, that works with ASPR staff on issues such as pharmaceutical shortages and supply chain needs. ASPR also runs the public–private Hospital Preparedness Program (HPP) that was started by the Health Resources and Services Administration in 2002. This program provides states with money to fund efforts by hospitals and local coalitions to prepare for disasters and develop surge capacity, hardened communication infrastructure, and relationships. To receive money from this program, hospital coalitions must work with their community response partners, such as emergency management, emergency medical services, and public health, on community preparedness. As the 2017 hurricane season showed, some of these coalitions are quite capable, while others have room for improvement, said Yeskey.

He then mentioned several other programs that aim to improve preparedness on a community or regional basis. In September 2016, the Centers for Medicare & Medicaid Services (CMS) finalized a rule to establish consistent emergency preparedness requirements for health care providers participating in Medicare and Medicaid, increase patient safety during emergencies, and establish a more coordinated response to natural and human-made disasters.[3] ASPR's Biomedical Advanced Research and Development Authority (BARDA), another public–private partnership, advances medical countermeasure candidates through the later stages of preclinical development and the initial stages of clinical development. Over 10 years, BARDA's efforts have helped develop 34 new countermeasures that have been approved by the Food and Drug Administration and are now included in the national stockpile for responding to chemical, biological, radiological, and nuclear (CBRN) threats.[4]

The NDMS that Yeskey mentioned earlier is a unique partnership among ASPR, DoD, the VA, Department of Homeland Security (DHS), and private-sector hospitals to create the capability of providing field care and moving patients out of harm's way to hospitals that the VA coordinates with DoD. This network includes some 1,900 hospitals out of the 5,000 U.S. hospitals. ASPR would like more hospitals to join the network and

---

[3] See https://www.cms.gov/Newsroom/MediaReleaseDatabase/Press-releases/2016-Press-releases-items/2016-09-08.html (accessed April 18, 2018).
[4] See https://www.phe.gov/about/BARDA/Pages/default.aspx (accessed May 1, 2018).

is looking at ways of getting more hospitals to engage with the program. Yeskey noted that during Hurricane Maria, the VA provided personnel to staff federal medical stations, made its clinics and hospital in Puerto Rico available to anyone, and opened its supply chain and offered logistical support to response efforts on the island. "That was new ground, and we look to strengthen those relationships and work with the VA on how to better do that," he said.

ASPR's priorities going forward, said Yeskey, are to provide strong leadership, enhance public health security, improve the medical countermeasure enterprise, and create a regional disaster health response system. In each case, ASPR will depend on strong, ongoing public–private partnerships. As an example of strong local leadership, he noted Texas's response to Hurricane Harvey in providing resources to move patients and establish situational awareness and support at the local level. He also singled out the nation's response to the 2017–2018 influenza outbreak and the way in which hospitals stepped up to take care of patients even when emergency departments were overwhelmed with cases.

In conclusion, he charged the workshop participants with helping to identify the respective roles of government and the private sector in preparedness and response and to call out best practices in places where public–private partnerships are having success. He also asked the participants to consider cases where the response was not optimal and to identify the barriers that prevented effective working relationships between government and the private sector. "We hear sometimes in ASPR that the federal government is not a good partner and that we do not hold up our end of the bargain sometimes," said Yeskey. "We want to know about those, and we want to understand what the challenges are so that we can get better at it." His hope for the workshop was that it would help define a clear, actionable path forward in the short and intermediate terms that will enable ASPR and the nation to address the challenges associated with 21st-century threats.

ORGANIZATION OF THE PROCEEDINGS

This Proceedings of a Workshop summarizes the discussions and panel presentations that took place during the workshop. Chapter 2 provides private health system and federal perspectives on the nation's capacity to respond to threats to health, safety, and security. Chapter 3 presents several examples of cross-sector collaboration from past disasters. Chapter 4 recounts small-group discussions about ASPR's new vision for a regional disaster health response system. Chapter 5 takes up the subject of how to cultivate best practices in disaster response at both the ground level and across the entire health care enterprise. Chapter 6 discusses regulatory and financial barriers and facilitators for engaging the private sector in building capacity to respond to disasters. Chapter 7 summarizes the closing keynote presentation by the Assistant Secretary for Preparedness and Response and the subsequent discussion with the workshop participants.

In accordance with the policies of the National Academies, the workshop did not attempt to establish any conclusions or recommendations about needs and future directions, focusing instead on issues identified by the speakers and workshop participants. In addition, the organizing committee's role was limited to planning the workshop. This Proceedings of a Workshop was prepared by the workshop rapporteur as a factual summary of what occurred at the workshop.

2

Perspectives on the Nation's Capacity to Respond to Threats to Health, Safety, and Security

The workshop's first two panels provided different perspectives on the nation's capacity to respond to threats to health, safety, and security. The first panel, moderated by Nicolette Louissaint from Healthcare Ready, featured four speakers who addressed the perspective of the private sector, which delivers the bulk of care in the United States. The four panelists were Brent James, formerly at Intermountain Healthcare and now at the Institute for Healthcare Improvement; Michael Wargo from the Hospital Corporation of America (HCA); Ronald Stewart from the University of Texas School of Medicine at San Antonio and the Southwest Texas Regional Advisory Council (STRAC); and David Witt from Kaiser Permanente. The second panel, which provided insights on the federal perspective, was moderated by Thomas Kirsch from the Uniformed Services University of the Health Sciences (USUHS). The four panelists were Kevin Hanretta from the VA; Melissa Harvey from ASPR; Anthony Macintyre from the Federal Emergency Management Agency (FEMA); and Jody Wireman from the U.S. Northern Command (USNORTHCOM) Headquarters.

PRIVATE HEALTH SYSTEM PERSPECTIVES

The first panel began with each panelist talking about the capabilities of their respective organizations. Wargo explained that as HCA's enterprise vice president of preparedness and response, he has oversight of nearly 200 hospitals and just below 2,000 outpatient facilities in 22 states. These are divided into 14 divisions that implement a framework of governance, operations, and tactics to deal with emergency operations and preparedness. Witt explained that Kaiser Permanente has, over the past 30 years, developed and drilled on a number of disaster scenarios and responded to several, including the explosive fire season that affected California in 2017.

James said that Intermountain Healthcare's preparedness activities began in conjunction with the 2002 Winter Olympics, which were held in Salt Lake City. That effort, which included a coordinated community response, focused in part on a potential bioterrorism attack. One result of that planning was that several hospitals in the region changed their physical infrastructure to be able to cordon off parts of the hospital in response to an infectious agent outbreak. Other outcomes included the development of GermWatch, an automated system for transmitting reportable diseases to the state health department, and a data system that can track local store supplies of over-the-counter medications to track epidemics, both of which are used today. James noted that the emphasis on community-wide coordination led to the development of hospital go-to designations for specific problems. "Frankly, dealing with this at a community level was dramatically easier than dealing with it at a facility level, and it seemed to make a real difference," said James.

From Ronald Stewart's perspective as chair of STRAC's executive committee, the regional trauma system serves as a great framework for disaster response, in large part because every health care system in the region—including emergency medical services (EMS), public and private health care, and public safety and public health—is included in the trauma system. The resulting diversity of facilities and capacities means the system is flexible and adaptable, said Stewart. Another advantage of this approach, he said, is that the trauma system, unlike a large health system, controls nothing but influences nearly everything, making it scalable, practical, and sustainable because the people who would respond to a large-scale event are the same ones working everyday tending to normal life events in their respective organizations. "I really do think it gets the right people, the right organizations, and the right leadership in the room together," said Stewart, "and when you get that group of people in a room together you can get creative problem solving."

Louissaint then asked the panelists to talk about how their organizations track events and determine what capabilities they will bring to bear in the subsequent response. Wargo explained that for situational awareness, HCA uses both internal and external communication systems, including off-the-shelf commercial products that scan social media such as Twitter as an initial indicator that something is happening in the community. If the number of tweets passes a threshold above normal, for example, his office then reaches out to HCA's divisions or regional advisory councils to gain additional intelligence. An internal incident management system provides real-time situational awareness across every HCA entity. "Situational awareness across the entire enterprise is critical to our operation so we can scale resources and mobilize from division to division or within a division so as to not stress any one facility or one area and to support those areas," said Wargo.

Kaiser Permanente's situational awareness activities, said Witt, began with preparations for the new millennium and when it was asked to prepare for a potential bioterrorism attack at the Rose Bowl. In response, Kaiser established a national group that addresses all threats and hazards and can garner and coordinate resources across the entire organization. One benefit that comes from this group is that representatives from each of Kaiser's regions meet regularly, enabling them to develop trusted relationships with one another. If any region has a concern, the group assesses it and assembles a team of experts that would be ready if an event develops. For Witt, a high point for this group has been its involvement in responding to influenza season, an annual large-scale disaster that kills tens of thousands of people and overloads every emergency department in the country. "We pretend this is a staffing issue, but it is really an infectious disease disaster, and we are finally using this mechanism to make sure that the command centers are open, and we are actually functioning in a disaster mode," said Witt.

James said he agreed with Stewart's statement that response systems work best when they involve those who work on a daily basis with a particular component of the system. He added, though, that coordination occurs more effectively within a big integrated system than outside of that system, particularly concerning communication. Stewart noted that communication breakdowns are the common thread in failures that occur during wide-scale events. His region has a consolidated communication system that coordinates some 400 hospital-to-hospital transfers annually among two military and one civilian trauma centers. This communication system is now funded by all health systems in the southwest Texas region. Today it also coordinates transfers of mental health patients between facilities. At times of an emergency, such as when there was a bombing the night before the workshop at a FedEx facility located between San Antonio and Austin, the center sends out notifications to the appropriate people and coordinates communication among the relevant parties needed to respond to the developing situation.

Calling the STRAC center an incredible model for coordinating and communicating, Louissaint then asked the panelists to talk about what works in coordinating activities by the public and private sectors. Wargo applauded the development of the Sector Coordinating Council, a public–private partnership operating under the auspices of the critical infrastructure branch of DHS and HHS. This council allows Wargo and others from the private sector to share information openly with colleagues in the public sector and gain both national and global situational awareness of potential threats. Communication is bidirectional, he stressed, in that it allows the private sector to inform government partners and leaders of what is happening in the private sector in the event of a large-scale emergency, such as hurricanes or the mass shooting in Las Vegas. "We are able to give real-time awareness to leadership, so they can scale appropriately and be more informed on the direct impacts," said Wargo. Another positive development, he said, has been the creation of community-based coalitions through the HHS-funded HPP that not only help communities to work together to ensure safe, uninterrupted care within communities in the face of a disaster, but also enable relationships across the various sectors in a community.

Witt noted that the structures that STRAC and Intermountain Healthcare have developed create a good framework for communication, but perhaps more importantly, for relationship building and cooperation among organizations that normally compete with one another in the private health care sector. In addition, said Witt, these frameworks create stability, institutional memory, and a cooperative culture that can withstand the inevitable turnover in personnel. One concern James has concerning Intermountain's communication system, though, is that 80 percent of the population in Utah lives along the geologically active Wasatch Front. In the event of a major earthquake, all of the region's communication capabilities would go down at the same time. Regarding the ability of regional councils to build strong relationships among private-sector health care organizations that are often intense competitors, Stewart said that these relationships help create needed redundancy in leadership that could come into play should the leadership of one system be knocked out of commission by a disaster.

Turning to the subject of gaps, Wargo said that information sharing among competitors is still a challenging prospect. Coalition models such as the regional advisory councils in Texas can help address that gap by serving as a consolidator of information on available beds and supply chain issues that would be closely held business intelligence, but critically important to have in the event of a large-scale disaster. He noted that the biggest challenges for HCA arose from the combination of Hurricanes Harvey and Irma, when communication among the various levels of government and the private sector was less than ideal. Too often, he said, there was conflicting information coming from boots on the ground and state and federal sources.

Witt said there are many issues concerning the interactions between the public and private sectors. For example, public health does not really deliver health care, but its surge plans call for taking over private health care facilities. "The one thing that would gum up our response would be someone appropriating our equipment and distributing it in a way they think is best," said Witt. In addition, while all organizations are risk averse to some extent, those in the public sector are particularly risk averse, which can be crippling at the time of a large-scale event. As an example, he said the delay in the Centers for Disease Control and Prevention (CDC) declaration that the H1N1 virus was not a novel agent likely cost the U.S. health care system $1 billion in wasted isolation equipment, created gridlock in hospitals, and caused harm to patients who were kept needlessly in isolation units. "We knew that this was not a novel virus for 8 months before that declaration was made," said Witt.

A step that is imperative, he said, is to figure out how the public sector can take some risks and accelerate its decision-making process. He suggested working with the private sector, mining its expertise, and creating memorandums of understanding to specify how the private sector would assume some of that risk. He noted that Kaiser participated in some exercises with the military during Fleet Week in San Francisco, and while the exercise was valuable, it became obvious that in the event of a giant disaster, 3 days would pass before the military could establish its medical facilities and aid in San Francisco's response. "There really are gaps between what the public sector believes it can do and what the private sector will have to do, and I think we need to work on those," said Witt.

James said the Utah State Department of Health and Utah Hospital Association used to serve as effective coordinators of regional activities in Utah. However, when he checked with his former colleagues prior to the workshop, he learned that the system is not as strong as it was when he was still at Intermountain Healthcare. The problem, as he understands it, is that the system has not been stressed for some time, and that the level of coordination needed requires active maintenance. That realization, he said, raises the issue of how to maintain a level of coordination and communication during good times.

Noting that STRAC served as the main coordinating center during Hurricane Harvey and the Sutherland Springs, Texas, church shooting that killed 26 people, Stewart said those two events stressed the system in different ways and revealed there was value in information sharing and coordination in general. The biggest gap those events revealed were in the coverage provided by the patchwork of trauma and emergency health care systems. In his mind, an affordable and feasible way to fill that gap would be to establish a set of minimum interoperable standards that would provide a framework for a trauma system designed intentionally for disaster response. This framework would include every public and private health care entity in a region.

Regarding his region, Stewart said he has noticed that private health care systems have become efficient, lean organizations over the past 20 years, which creates problems during a wide-scale event because each organization will have little excess capacity. "But with a diverse, inclusive system [with components that are] sharing information with each other, while one facility may have very limited surge capacity, you can distribute that across the entire system and balance that load," said Stewart. He noted that during Hurricane Harvey, health systems preferred to transfer patients among their own facilities, but STRAC was prepared to help find capacity outside of those systems when it was needed. "The critical issue during a widespread event is bed capacity, but that can be dealt with at least in part by improved coordination and sharing across health systems," he said.

For Wargo, a major concern is the potential for cyber threats to take down utilities in a region. He believes the health care system has not mapped its interdependencies as thoroughly as it should to understand the long-term impact of losing the electrical grid or water supply. The saline shortage that occurred during Hurricane Maria is a case in point, he said.

Louissaint, noting that most health care in the United States is provided by the private sector, asked the panelists to comment on who they believe should be responsible for issues of health and security during a large-scale regional or national event. "I think everyone is responsible, even to the level of the single community member," said Wargo, referring to individuals providing nuggets of information about what is happening on the ground. When it comes to the response and coordination aspects, Wargo said, the initial response needs to come from the local community, which needs to establish who has command and control authority. He noted that each organization, public and private, brings different resources and capabilities to the response, and it is important to establish a scalable chain of command and information pathways to coordinate how each partner can contribute to the response. Unfortunately, he added, during the recent hurricane season, FEMA had one approach, HHS had another, and DoD brought a separate set of capacities to the table. The private sector, meanwhile, did not have the information to understand the whole of the community and public-sector response, and thus, was slow to brings its resources to the response effort.

In Witt's opinion, everyone in the public sector and private sector knows they are responsible for doing something, but the key is having one entity coordinating those responsibilities. He has a genuine concern about the medical community not accepting its responsibility for planning and training. Much of the training, he said, has been developed for prehospital care, yet the core response to most disasters will be to provide hospital-based or urgent care. James agreed that coordination is key to bringing together the different scales of responsibility to produce an effective and efficient response. His concern relates to the occasional conflicting messages he has received from those who think they are in authority and the potential damage that can result from conflicting messages.

Stewart agreed that everyone is responsible in a disaster, but having a framework to facilitate that responsibility helps. He noted that preparation is key for responding to wide-scale events, but with most private health care systems it is difficult to get leaders' attention to prepare for a low-probability event, even if the potential impact is high. "Having a framework that facilitates preparation is key," said Stewart, who acknowledged that public health, law enforcement, and the fire department are better able to respond to a disaster because they have a structure that allows for training and preparation. Health care, in general, lacks such structures, so having a framework that gets public safety, health care, and public health working together through the regional trauma system can create a trauma system that serves as a regional disaster health response system. He said the Texas regional trauma center, which operates every day at the local level, provides such a framework. A regional not-for-profit organization has contractual obligations to the state to coordinate activities to improve the system, Stewart added. HCA, said Wargo, has added leadership that is responsible and accountable for readiness, response, and recovery. Witt noted that he is responsible for ensuring that Kaiser's operations in northern California will be able to continue to deliver services during and after a disaster. He added that government can best serve as a facilitator or a responsible party, rather than a director.

DISCUSSION

Ricardo Martinez from Adeptus Health started the open discussion by asking the panelists to comment on how the private sector can interface with public health when the response to a regional disaster, such as Hurricane Harvey, has to transition from an acute care mode to one that has to provide care for someone who is homebound and cannot get an oxygen delivery, for example. HCA, said Wargo, has a hurricane playbook that starts 120 hours before the hurricane, continues through the hurricane, and turns to community resilience once the immediate event has passed. This playbook accounts for whether the HCA system is intact or has structural damage that alters its ability to care for its community, and if so, how it can mobilize clinical resources or equipment from other parts of its system to address the needs of the affected community. In the case of Hurricane Harvey, where HCA's East Houston Medical Center was flooded and remains closed, the organization had to turn to its partnerships through the regional councils to understand where its community members would migrate for their care and how it could draw internal resources to deliver care. "Sharing risk and sharing impact across the community versus burdening any one individual health system is the approach that we take," he said. Wargo added that HCA, in partnership with NDMS, had worked out ahead of time that its hospitals, augmented with state and federal resources, would serve as triage centers during Hurricane Harvey. The key point was that this plan was made in advance, rather than having the federal or state government commandeer HCA's facilities and making a game plan on the fly.

Witt said that Kaiser, as a fully integrated health system, has an easier task in those situations because it can draw on its home hospice program, for example, to deliver services in individuals' homes rather than stress the capacity of the hospitals and emergency departments. Every hospital also has connections to coordinate and work with home health organizations to expand home health care capacity in such situations. James added that an integrated health system, such as Intermountain or Kaiser, can also draw on all of the components of the system, such as walk-in primary care facilities, in its response. His advice was to consider the entire system, not just the hospitals, in planning for a long-term adverse situation.

Craig Vanderwagen from East West Protection noted the tension he had heard from the panelists between network coordination and hierarchical failure, and he asked the panelists how the public sector could be more effective in supporting network thinking given that the public sector is largely driven by hierarchical thinking. Stewart replied that the first step would be for the public sector to accept the importance of the network and the system as a whole. The Texas trauma system, for example, is built on a professional model of evidence-based self-government rather than on a regulatory framework. "I have professional freedom, but I have the responsibility to do the right thing for the patient," he explained. His advice would be for the federal government to create a set of principles and minimum standards—with input from the private professional community of doctors, nurses, paramedics, and others—that it would use to facilitate the formation and activities of regional systems.

James noted that networks rely on information transfer, so government agencies should explore what they can do to improve information transfer. Wargo suggested that the federal government could develop a standard framework to address compliance issues related to moving staff across geopolitical borders during a crisis. During the 2017 hurricane season, some entities were applying for 1135 waivers[1] in a coordinated manner, while others were not. Having a mechanism in place to execute those waivers quickly would accelerate resilience, he said.

Lewis Kaplan from the University of Pennsylvania Perelman School of Medicine wondered if large health care systems should have a single disaster preparedness dashboard that everyone in a region can access and that can interface with government resources. He also asked if hospitals should be viewed as sites of expertise that can be exported to the public health system to create a durable and embedded link for distributed competence

---

[1] See https://www.cms.gov/Medicare/Provider-Enrollment-and-Certification/SurveyCert EmergPrep/1135-Waivers.html (accessed April 19, 2018).

during a prolonged emergency. Witt replied yes to both questions and noted that Kaiser recently adopted an electronic emergency department visibility system for all its hospitals. "That is just within our system, but it would be great to have for a region," said Witt. Wargo said that Pennsylvania has an electronic system that could provide that capability, but it is only updated monthly or so. He would like to see the establishment of regional advisory councils that would oversee an interoperable system where health systems would deposit information in real time. In fact, he said he has called for the creation of standards and requirements for interoperability across dashboard systems.

John Hick from the Hennepin County (Minnesota) Medical Center asked the panelists for their ideas on how to balance the tension between the ability to standardize and synthesize across different systems and the ability of an individual system to create its own standard workflow. Wargo said that HCA's number one mandate to its facilities is to integrate locally first. For example, the local coalitions have standardized which powered air-purifying respirators they will use, and HCA's warehouses now have a cache of that equipment. "You have got to have local integration in whatever the framework," said Wargo, "but if we are looking at a national model, having a national standard is a start so that we can then take the disparities across all our coalitions and maybe bring it down into a tighter framework so we have more interoperability."

Stewart agreed that local integration followed by regional integration is the right approach, with the federal government providing some standards to encourage interoperability, particularly regarding electronic health systems and dashboards. He added that in the Texas system, there is a mindset of health systems learning from one another and working together to focus on what is needed when the next big disaster occurs.

Duane Caneva from the National Security Council noted the fragmentation of the medical system and called for a matrixed approach that would capture local, regional, and national standards as well as the interdependencies across critical sectors. Disasters, he said, are not just one-time events because the threats are ongoing, including the annual influenza disaster that Witt discussed earlier. "What is the organizing structure at the national level that we can look to that allows the house of medicine to become a house unified?" asked Caneva. Witt replied that there are structures that exist yet are problematic, such as electronic health record systems that do not talk to each other, even though they should and could if there were national standards and defined elements. Wargo suggested that the elements exist, but the organization does not. "I think we need a clearly defined mission structure of where we want to start tackling that," said Wargo. Stewart said that if the various sectors and professional societies committed themselves to approach things from a professional, evidence-based self-governance model, with a commitment to civility, collegiality, and professionalism, standards and frameworks for interoperability could be developed.

Brendan Carr from ASPR remarked that a central theme of the panelists was to build on functional day-to-day systems to create a larger disaster-ready system. Given that, he wondered if there is a proxy that can provide insights about which parts of the day-to-day systems would be best able to respond when the system is pushed to its maximum capability. As an example, he noted the way that hospital readmission rates and hospital-acquired infection rates have served as proxies to identify quality improvement priorities for health care systems. James replied that one mistake at the national level has been to define standards from the top-down for an immature industry, as that suppresses innovation in industries that need innovation and advancement. This was the case, he said, when the meaningful-use standards were issued, which had a chilling effect on electronic health record innovation. He suggested a focus on transparency in terms of making information available to those who need to execute using that information when it is needed. He also pointed out that building something at the systems level so that it does not require constant attention helps with sustainability and makes it easier for people to do the right thing.

Stewart said that if he was informing federal policy, he would not recommend competing around quality, because quality should be what everyone is enabled and encouraged to do. In that regard, pay for performance, in his opinion, encourages health systems and providers to meet a metric, which he believes is a major distraction. Instead, he would encourage the process of improvement by setting standards that raise the quality of care and ensure development of the right infrastructure.

John Dreyzehner from the Tennessee Department of Health noted that the nation's preparedness for a predictable disaster—the annual influenza season—is designed to fail given that the nation only makes about half of the vaccine needed to immunize everyone in the country. Given that, he asked the panelists how they would redesign the current system to prevent the majority of influenza-related deaths and illnesses. Witt replied that he wished he had the answer to that problem, though he noted that the rate of influenza vaccination has improved dramatically in the United States. One problem, he said, is that the nation charges for influenza vaccine, which is a barrier for some people. Another problem is that the health care system does not require all employees to be vaccinated. "There are conflicting issues of autonomy versus society benefit that have not been solved," he said.

For the final question of the session, Sara Roszak from the National Association of Chain Drug Stores asked the panelists if they had ideas for quality metrics that could be used for preparedness purposes. The problem, replied Witt, is identifying what a metric in emergency preparedness would be. Metrics would have value, he said, but it will be important to identify the unintended consequences of a metric before implementing it.

FEDERAL PERSPECTIVES

Before having the members of the workshop's second panel describe their agencies' activities, moderator Thomas Kirsch said that from his perspective as a first responder, emergency physician, and director of the National Center for Disaster Medicine and Public Health, engaging the private sector is critical for preparing the nation to respond to large-scale disasters. Moving on to the panelists, Keven Hanretta explained that the VA is the second largest department in the federal government, with more than 374,000 full-time employees, nearly 200,000 contractors, and 100,000 volunteers and affiliates. They support the health care of more than 9 million veterans at 1,700 points of care across the United States. In the event of a disaster, the VA expands its responsibility to deliver care to the 19 million living veterans in the United States.

Given that all disasters are local, the local VA facilities are often just as affected as private health systems, so its first responsibility following an event is to stabilize the VA facility so that it can continue to provide health care for its veterans, Hanretta said. At the same time, FEMA can call on the VA to provide support to state and local governments, as can HHS as part of NDMS. In fact, Hanretta explained, the VA provides 50 of the 63 NDMS coordinating centers spread across the nation. These coordinating centers provide the staff who coordinate the reception of patients coming out of a disaster area into civilian hospitals. The VA also has the congressionally mandated responsibility to be DoD's contingency hospital system in the event that DoD evacuates casualties from the battlefield and requires surge capacity. "Whether it is FEMA asking us, HHS asking us, or DoD asking us, [the] VA has that responsibility to be a national asset, to step up and be able to share the resources that we have across the United States," said Hanretta. In that regard, he added, the VA is involved in all disasters that occur in the United States.

Melissa Harvey then described ASPR's HPP, which focuses on engaging the private sector through a cooperative agreement grant program currently funded at $255 million per year. Since 2002, in response to the 9/11 and anthrax attacks, these funds have gone to the health departments in every state, territory, and freely associated states, as well as the District of Columbia, New York City, Chicago, and Los Angeles, to build capacity and capabilities among the largely private U.S. health care system. While the program's initial efforts were directed at building surge capacity at individual hospitals, it has expanded to build capacity across regional health care coalitions after recognizing that hospitals will be overwhelmed during an emergency unless other components of the health care system cannot be somewhat resilient on their own and be integrated into regional response plans.

Currently, 470 health care coalitions, composed of some 31,000 member organizations, participate in HPP, but Harvey expects that number to drop given a new requirement that every HPP-funded coalition must include four core members, of which two must be hospitals, as well as local emergency management and local public health. Though the number of coalitions will fall, largely through the merger of smaller coalitions to meet the new requirement, the number of member organizations is expected to increase substantially because of the new CMS Emergency Preparedness Rule that went into effect in November 2016.[2] This rule, explained Harvey, applies to nearly 70,000 different providers and suppliers, and establishes a baseline of preparedness for those individual health care entities. Hospitals will already meet these standards because they are the same as those required for accreditation by the Joint Commission, but these standards will be new for many outpatient providers, dialysis centers, urgent care centers, and other types of facilities. "Emergency preparedness in some cases is completely new to them, and that's why we think that this is a huge step forward," said Harvey. She noted, though, that even with these new baseline standards, the challenge will be to engage the executive leadership in health care in preparedness, readiness, and the importance of joining these coalitions.

FEMA, said Macintyre, is best known for its role in coordinating interagency relief efforts for presidentially declared disasters and emergencies and for administering the Disaster Relief Fund as outlined in the Stafford Disaster Relief and Emergency Assistance Act.[3] As he explained, the Stafford Act contains specific details about how FEMA and other federal agencies can engage and support regions affected by these presidentially declared incidents. He noted that FEMA, working through specific federal agencies such as HHS, provides assistance that the affected state, territory, or tribal government has requested or prioritized, though that can play out in unexpected ways. As an example, Macintyre recounted one incident following a recent hurricane in which the governor of the affected state requested assistance, but when his search and rescue team arrived at one town, the sheriff and mayor stood at the bridge leading into town and told them they did not want any assistance from the federal government.

In addition to its Stafford Act roles, FEMA can be called on to coordi-

---

[2] See https://www.cms.gov/Medicare/Provider-Enrollment-and-Certification/SurveyCert EmergPrep/Emergency-Prep-Rule.html (accessed April 20, 2018).
[3] See https://www.fema.gov/media-library/assets/documents/15271 (accessed April 20, 2018).

nate other interagency responses to other crises. Examples include the 2014 presidentially declared unaccompanied children humanitarian crisis along the U.S.–Mexican border and the HHS-led response to the Zika outbreak in Puerto Rico. Macintyre explained that HHS, not FEMA, is the primary federal entity responsible for health and medical preparedness, but FEMA does work with HHS on multiple initiatives, such as the Biological Incident Annex, which outlines the actions, roles, and responsibilities associated with response to a human disease outbreak of known or unknown origin requiring federal assistance.[4]

The North American Aerospace Defense Command (NORAD), explained Jody Wireman, director of the Force Health Protection Division, is part of USNORTHCOM, and therefore it serves as the DoD operational lead for events that occur in the United States, and in particular, for those that fall under a FEMA mission assignment to support local, state, and regional authorities. USNORTHCOM's response capabilities are divided into chemical, biological, radiological, and nuclear and other kinds of events such as hurricanes and earthquakes. The CBRN response enterprise includes 27 National Guard response groups and one large DoD response group, the latter of which includes four Army and two Air Force units. Wireman noted while responses to hurricanes and earthquakes typically take longer than a CBRN response, DoD medical units do respond once the danger of a subsequent CBRN event has passed. During the 2017 hurricane season, DoD mounted its largest response ever to an event in the United States, she said.

Wireman explained that USNORTHCOM's medical capabilities are under the command of the individual services, and local-level support would be at the installation level. In some regions, such as southwestern Texas, the trauma capabilities of the local military facilities are well integrated with civilian capabilities. In other regions, the opportunity exists to form better relationships that would benefit both local capacity and DoD operations, she said. Across the United States, USNORTHCOM relies heavily on regional planning efforts, and would welcome the opportunity to be more engaged with and integrated into those planning activities.

With the introductions complete, Kirsch asked the panelists to talk about how their agencies were involved in the responses to the Ebola epidemic and the 2017 hurricane season. Regarding the hurricanes, Harvey said one of the big lessons was that each coalition is unique in its operations. The Texas coalitions, she said, were built without HPP funds on top of an established, high-functioning trauma system, and they did an excellent job during Hurricane Harvey. The Houston-area coalition has a physical

---

[4] See https://www.fema.gov/media-library/assets/documents/25550# (accessed April 20, 2018).

location for its Catastrophic Medical Operations Center that coordinated information sharing and did a good job orchestrating the task of transferring patients from affected hospitals to those sites that could receive patients. When a regional psychiatric residential treatment facility, which was able to shelter its patients in place, began running out of the medications their patients needed, the operations center worked with area hospitals to get those medications and with the volunteer group of boat owners that called themselves the Cajun Navy to pick up the medications and deliver them to that residential facility. That type of coordinated response, said Harvey, kept the limited number of mental health beds in hospitals from being overwhelmed.

This type of exceptional response is not something that occurs with all 470 coalitions, many of which are either too small or too dispersed geographically to effectively plan, let alone operationalize, a disaster response scenario. Too often, she said, coalition members do little more than exchange business cards and then revert to their individual roles in their own individual health care facilities when an emergency takes place. "Those coalitions have to stand up and be able to share information and coordinate resources," said Harvey.

Harvey noted that ASPR received some 1,800 comments from coalitions who responded to the new requirements, and the clear majority said that health care coalitions are planning bodies only, not response entities. "I don't know about you, but I do not know what you are planning for if you are not going to respond," she said. Ironically, the physicians at those facilities all consider themselves to be frontline responders, which in Harvey's view makes their facilities frontline responders. "That means we have to make sure the coalitions view themselves as having a role," she said. In some cases, the coalitions are designated to vet resources, move them around, and serve as logistical coordinator. When that is not the case, it is still important in ASPR's view that these health care systems have a good situational awareness and an idea of what resources they can bring during an event.

During the Ebola outbreak, surge capacity was not an issue, but surge capability was, in the sense that the dozen or so patients brought to the United States taxed the infectious disease capabilities of the health care system to their maximum. What was interesting about that event, said Harvey, was that never in HPP's history had there been such a high level of executive engagement around an emergency. "If we could figure out a way to bottle up that engagement and that energy around what was an event that really impacted only a dozen patients, I really think that is where we can make some major progress," she said. As a final thought, Harvey said that while she and her colleagues at ASPR talk about how coalitions need to be based on a good trauma model, Ebola was not an event that depended on a well-developed trauma system. In her opinion, that fact points to the need to take a systems approach, rather than just a trauma-based approach, to be ready to handle mass shootings or infectious disease outbreaks.

Macintyre noted that FEMA is still undergoing its after-action exercise from the 2017 hurricane season, but in his opinion, there is more work to be done to improve private-sector resiliency even with the considerable progress that has occurred with the development of best practices gained from real experiences. As an example, the private-sector dialysis system in Puerto Rico was tremendously well organized and had generators ready for most of the clinics there. "Clearly, they still needed support after the storm, but if those pieces had not been in place, we would have been looking at an exodus of somewhere on the order of 6,000 people within 48 to 72 hours who needed dialysis," said Macintyre. Where there is room for improvement is in what he called big-ticket items that involve large capital outlays. He recounted an effort that DHS led in the National Capital Region nearly a decade ago that examined the infrastructure of every hospital in the region and developed recommendations for how to fortify water and power systems, for example. At the end of the day, however, there was no funding to make the recommended improvements. In his opinion, the nation needs to develop some innovative methods for funding those types of improvements.

Other places where Macintyre sees room for improvement are in information management and sharing and in hardening communication facilities. He also noted that the trend to provide more care outside the hospital means there will be a large population of people receiving care outside the health care system who will need support during a disaster. "I know ASPR is paying attention to this, and FEMA is certainly paying attention to this, but it is going to become more of a problem," he said. "We certainly saw challenges with this population during the last hurricanes."

When responding to the 2015 earthquake that devastated Nepal, Macintyre and a colleague had a wonderful view of the international efforts that helped the country respond and recover. One asset, though, proved to be particularly vital to the medical response—NGOs that focused on rehabilitation and were designed to get people out of hospitals. "Having those organizations that could decompress the acute care facilities really saved the day, so that is an area for us to focus on," he said.

The mantra for DoD in engaging during a disaster, said Wireman, is last ones in, first ones out to avoid getting in the way of what local, state, and regional entities are doing and to make sure that assets are available when requested by the local authorities. DoD's response in Puerto Rico was challenging, she said, because while it had teams ready to go, it was difficult to integrate information coming from the island about prioritized requirements for food and water with the island's medical needs. Another challenge was that DoD doctrine calls for sending out its "platform tonnage" along with its people, and that can take a long time to deploy, raising the question of whether it would be possible to just send medical teams and just the necessary equipment in future responses. "I do not think we have explored that well enough," said Wireman.

At the same time, while the military response can be slow, the delay offers the opportunity to accurately assess needs following the first 72 to 96 hours of response, when the initial responders may be at the point of exhaustion and stockpiled supplies are being drawn down. Wireman noted that DoD can deploy worker health and safety people sooner, and in the case of Puerto Rico, DoD was able to work with the Puerto Rican Health Department on water quality and mosquito trapping.

For the Ebola response, Wireman said she is not sure the U.S. government or even DoD realizes how integrated DoD was with the public and private responses to the outbreak. In fact, DoD looks to established private and public facilities as the first choice of where DoD members would go if they were infected with the virus. Even for training, there were efforts by the individual services to link with the University of Nebraska to supplement the limited capacity of the Army's Institute for Infectious Disease. "In many ways, we look to the public and private sectors to gain information that we can incorporate into our processes," said Wireman. In the case of Ebola, that meant using the regional mechanism that HPP has established for Ebola and other biological agents.

DISCUSSION

Harvey Ball from the Administration for Children and Families asked the panelists if anyone has looked at whether HPP dollars save the government money by preventing the deployment of federal assets, which can be expensive. Harvey replied that ASPR has just started that analysis and added that nobody believes that grants can fund all the nation's preparedness needs. "Demonstrating return on investment is important, but there needs to be a recognition that these programs need to be sustained and moved further along with state and local support," Harvey said. Macintyre added that much of what happens at the federal level helps engender and build systems at the state and local levels in ways that are tangible, but difficult to measure. For example, HPP, by funding coalitions, is building systems that help address everyday emergencies in a more efficient manner, he said.

Kaplan, noting that the coalition system is spread unevenly across the nation, suggested that ASPR should leverage the computing capabilities of the U.S. National Laboratories to ask the question of where coalitions are needed most and overlay that with the existing map of trauma centers. That would enable matching infrastructure that already exists to support new coalition activity. Doing that, said Kaplan, could open the door to conversations with health system leadership that might engage them in preparedness. Harvey thanked Kaplan for that suggestion. She pointed out, however, that rural areas in particular have little capacity and yet in some ways need the type of coordinating capabilities a coalition would offer

Noting the absence of someone from CMS on the federal perspectives panel, Daniel Hanfling from the Johns Hopkins Bloomberg School of Public Health asked if there was a role for the federal government, perhaps through CMS, to incentivize health systems to build facilities that would be more resilient in the face of a natural disaster. Macintyre replied that some guidance already exists in the safe hospitals program that the Pan American Health Organization (PAHO) and the World Health Organization have espoused. He acknowledged that health care systems could pay more attention to facility resilience and perhaps less to architectural beauty when planning new facilities.

Eileen Bulger from the University of Washington and the American College of Surgeon's Committee on Trauma asked the panelists for their thoughts on how to encourage a relationship between health care coalitions and existing, high-functioning trauma systems in the way that Texas has and to leverage health care coalitions to strengthen day-to-day responses in rural areas or regions where there are gaps in trauma capabilities. Harvey cautioned not to leverage the coalitions in areas where the trauma systems are not very strong because the way those systems were developed is not necessarily going to lend themselves well to day-to-day operations. In those cases, she said, it may be necessary for ASPR to work with organizations such as the American College of Surgeons and local trauma leaders to build a trauma system that can then serve as a foundation for a coalition.

Bulger then commented that the American College of Surgeons is working closely with DoD on integrating and strengthening military and civilian capabilities for local trauma systems. That type of integration proved to be important during the Las Vegas mass shooting event, when military personnel were allowed to help in civilian trauma centers. Harvey agreed that type of cooperation should play a role in strengthening local resiliency for disaster response.

Dreyzehner asked about the VA and DoD positions on vaccinating their health care workers against influenza and how they provide vaccine for veterans and active duty personnel. Hanretta replied that the VA takes the possibility of an influenza pandemic very seriously. It stressed its vaccination program every year and achieves a vaccination rate among its employees of about 64 percent. While the VA cannot mandate vaccination, the agency for the first time did not deploy VA employees who were not vaccinated during Hurricane Harvey. Regarding the veterans under its care, the VA vaccinated some 2.5 million individuals, which does not count the ones its pays for when they are vaccinated at their local drug stores.

Wireman said all members of the military are required to get their annual influenza vaccination. What he said he would be interested in is whether it would be possible to set aside funding for some of the public–private entities supported by BARDA to produce vaccine, particularly in years when the initial designation of the target viral strains was wrong. DoD also has a public–private pharmaceutical initiative that might be able to support such a program.

Hick asked the panelists for their ideas on how the nation could do a better job with medical intelligence and resource matching. Harvey replied that information sharing and management is something that could be done better at the federal level and that the coalitions are doing a better job getting information from their members using various information technology platforms, though by the time that information trickles up to the federal level that information is too old to be of much value. One problem in addressing this challenge is that the federal government has not clearly defined what it needs to know and what the benefit to the private sector would be for sharing that information. "There has got to be a value proposition, and that is something we need to begin to tackle," said Harvey. Wireman added that USNORTHCOM, the Transportation Command, and HHS are working to have teams that coordinate information dissemination, and Hanretta noted that the VA needs more help with both communication technology and telehealth. Better telehealth capability would have made a tremendous difference in Puerto Rico, he said. In a final remark to close the session, Mashid Abir from the University of Michigan Medical School wondered if the VA and private hospitals could share intensive care unit and burn care capabilities in a bidirectional manner.

3

Learning from Experience

The workshop's third panel session presented some lessons learned from the past that might guide subsequent discussions on what form public–private collaborations should take. Ricardo Martinez from Adeptus Health moderated this session, which included presentations by Scott Cormier from Medxcel Facilities Management; Karen DeSalvo from the University of Texas and formerly from the New Orleans Health Department; Erin Erb from the Gulf Coast Division of the Hospital Corporation of America; Todd Sklamberg from Sunrise Hospital and Medical Center; and Richard Zuschlag from Acadian Ambulance Service. An open discussion followed the five presentations.

CASE STUDIES IN CROSS-SECTOR COLLABORATION FROM PAST DISRUPTIONS

Scott Cormier began his presentation by explaining that Medxcel is part of Ascension Health Care, the largest not-for-profit health care system in the country. He mentioned this because large health systems such as Ascension, HCA, Kaiser, and the VA have a great deal of experience in emergency management. "If you have one disaster a year at your hospital, and that is a lot, in 10 years you will go through 10 disasters," said Cormier. In that same 10-year period, Ascension will go through about 1,500 disasters. "Last year," he noted, "we had 96 incident command activations, and it is this experience that leads the way health care systems respond to disasters." In fact, added Cormier, he has asked the American Public Health Association to bring the large health systems together to pool their knowledge, which he believes would have a dramatic effect on large-scale disaster response. Besides knowledge and experience, large health systems are self-sufficient, which means they will be less likely to draw on resources from the community and should be able to contribute resources and expertise to their communities.

According to Cormier, Ascension's emergency management program is unique in that it combines safety and emergency management as a means of bringing value to emergency management and spare it the occasional budget cuts that can happen when there have not been any disasters in the immediate past. He explained that all of Ascension's safety and emergency management people are under his chain of command and that there is a virtual emergency operation center he and his staff use to manage disasters from the very beginning. "It is not the typical model of 'good luck, God bless, hurricane's coming, if you need help, call us,'" he said. Rather, his group is part of the planning process at every Ascension facility, and the organization has national contracts with meteorological companies, generator companies, and supplies so that the organization is a self-sufficient entity when disasters strike.

He and his regional directors also stay in regular contact with state and federal partners, creating established two-way information conduits for when the need arises, as well as with Ascension's senior executive, both in times of calm and during a major event. "Once or twice a day during a major event, we will send senior management a very high-level update about what is going on, and they trust us to manage the incident and do the right thing," said Cormier.

Ascension has 12 hospitals in Texas, mostly located in the Austin area, and while there was some flooding and water damage during Hurricane Harvey, those hospitals served as receiving facilities for people evacuated from southeastern Texas. He noted that he was in contact with his friend and colleague Michael Wargo from HCA throughout Hurricane Harvey to make sure the two systems coordinated their activities. "I call that muscle memory communication," said Cormier, who joked that half of his salary pays him for his extensive contact list. "Being able to make those phone calls and share information or get resources is crucial. That is how we solve the communication problem during disasters, and we need to develop better muscle memory communication," he said.

One of Ascension's resources is its subscription to a private meteorological service whose models incorporate the physical location of all of Ascension's hospitals and some of its secondary sites. As a result, the weather reports it receives are detailed to the point that they can pinpoint the time and location of weather events that could affect specific locations in the Ascension system. This tailored and accurate information was crucial during Hurricane Harvey because the local and national weather reports were more spectacular than useful. "Emergency management becomes the source of truth during a disaster," said Cormier. "We saw that with H1N1 [influenza], we saw it with Ebola, and we saw it with the hurricanes. People depend on us to be that single source of proper information. We see that as a key role, so we make sure we are communicating that information regularly."

As an example, he recounted how during Hurricane Irma, the Jacksonville Fire Department came to one of Ascension's nursing homes and said it was there to evacuate the facility because reports were calling for the St. Johns River to rise 5 feet, which would have devastated the nursing home. Cormier called the local incident commander and told him that he had reports that projected that the river would rise no more than 30 inches and that he was prepared for that event. After some back and forth, the incident commander deferred to Cormier's better information. In the end, the river rose 28 inches. The same type of situations arose during the 2009 H1N1 influenza pandemic and 2014 Ebola outbreak, when the state health departments across the country were asking Ascension to do things outside of its scope of what CDC was recommending and what his system's best practices were. "We said no, and that is an important part of our program," said Cormier. "When you are the holder of the truth, that is what you need to do."

Cormier noted that the federal government was extremely helpful during Hurricane Irma by supplying platelets, which enabled Ascension's three Jacksonville hospitals to keep its surgical facilities open and operational. ASPR held daily conference calls during each of the national emergencies in 2017 through the Health Care and Public Health Sector Coordinating Council.[1] Cormier stressed the importance of being a member of the council. His team also worked with the Federal Bureau of Investigation (FBI) during the Austin bombings and is a member of a workgroup formed in 2013 to provide guidance on how health care systems can plan for and respond during an active shooter incident.[2]

Going forward, he said, one area that needs more work is how to leverage private-sector expertise with a federal response more effectively. As an example, he noted that his organization has expertise on how to restore a hospital after a hurricane, and yet nobody from Ascension is in Puerto Rico helping the island restore its hospitals. "Why is the private sector excluded from the Center for Homeland Defense and Security master's program?" he asked, noting that only government employees can participate. "Wouldn't it

---

[1] See https://www.phe.gov/Preparedness/planning/cip/Pages/partner.aspx (accessed April 23, 2018).
[2] See https://www.fbi.gov/file-repository/active_shooter_planning_and_response_in_a_healthcare_setting.pdf/view (accessed April 23, 2018).

be great to share our information with the classmates and learn what is going on in the government sector?" Another area that requires more thought, he said in closing, was mental health as an emergency management problem given the preponderance of mass shootings.

New Orleans's Katrina Experience

In 2005, Hurricane Katrina became the costliest storm in the nation's history, killing some 2,000 people and destroying infrastructure across the Gulf Coast. As Karen DeSalvo noted, the hurricane missed New Orleans, but the city's flood wall system failed and 80 percent of the city—a land mass equal to the size of Manhattan—was flooded. In the flood's aftermath, New Orleans was in a mandatory evacuation scenario for 30 days, and the city's entire health care, public health, and emergency response systems were severely compromised. DeSalvo explained that despite the evacuation order, tens of thousands of people remained in the city.

In the days immediately after Hurricane Katrina, no resources were available to provide medical care. The private sector rushed into that void to establish a set of makeshift sites around the impacted areas of New Orleans, as well as across affected areas along the Gulf Coast. For example, Tulane Medical School, where DeSalvo was on the faculty at the time and which was an HCA facility, set up urgent care stations where volunteers could give tetanus shots, provide emotional support to people coming to these stations, and fill prescriptions (DeSalvo, 2005). Most of the federal and local resources that eventually came into play were stood up in places such as Baton Rouge and Houston, where the bulk of the people had evacuated, she said, adding that DoD proved to be an extraordinary resource during this catastrophe, erecting portable field hospitals.

In the months after Hurricane Katrina, health care officials in New Orleans began thinking about how to build a more resilient system. One realization was that New Orleans had a very centralized health care system, said DeSalvo, with about 30 percent of the city's population receiving care at the city's Charity Hospital, which was also the city's level one trauma center, a major training ground for providers in the community, and an important source for outpatient care (DeSalvo, 2006). "We did not want to be back in a situation where when one hospital flooded, we are pretty well knocked out of the game, even with private hospitals able to stand back up more quickly," said DeSalvo. The remedy for New Orleans was to create a more distributed network[3] grounded in public health (DeSalvo and Kertesz, 2007). The city also worked to change financing so that people had portable insurance coverage that would enable them to receive care in other

---

[3] See http://504healthnet.org (accessed April 23, 2018).

communities, and it digitized the care experience so that providers would not be left guessing about people's medical histories.

Thirteen years later, she said, New Orleans has built a more resilient health care system with these characteristics (DeSalvo, 2016, 2018). In addition, it has also reimagined how to build health care facilities that would be more resilient in the face of natural disasters, something for which HHS has since developed a toolkit (Guenther and Balbus, 2014). Hurricane Katrina also led to national changes, such as taking an all-hazards approach to planning and conducting drills. According to DeSalvo, the years after Hurricane Katrina have also moved the emergency preparedness field to engage in more cross-sector and agency-wide planning, and the federal government to establish a strong role and set of federal resources, including an incident command structure. The health care sector also moved from a reactive position to being more proactive not only with regard to planning, but also knowing that those who planned would be responding in the field. "There is an old adage in internal medicine that discharge planning begins at admission," said DeSalvo, and New Orleans adopted that philosophy when it came to rebuilding its health care infrastructure and making it more resilient by design, which included making the entire community more resilient in the face of disaster.

One important lesson learned from the Gulf Coast's experience with Hurricane Katrina, and one DeSalvo said was reinforced during the 2017 hurricane season, was that the social determinants of health, which are often affected severely in a disaster, have a disproportionately negative effect on communities of color, people with low income, and seniors. As a result, she said, "We cannot just pay attention to making certain they have good access to good medical care. We have to attend to the other infrastructure that will impact their lives, since so many are living on the edge every day."

Human capital and capabilities cannot address all of those needs, said DeSalvo. "It is one of the reasons that we have as a country, and certainly in New Orleans, wanted to leverage data and technology to improve the effectiveness of response and recovery and resiliency," she said. In addition to the widespread adoption of electronic health records, the past 13 years have led to a significant shift in how health care data are used during a disaster. For example, when she was the New Orleans health commissioner in the years after Hurricane Katrina, she led an effort to leverage claims data from Medicare to develop a tool called emPOWER. This tool, which ASPR and CMS now manage, allows health care to target the most vulnerable seniors and other Medicare beneficiaries in the community who will need special attention in a disaster.

DeSalvo noted that health care systems spend a great deal of time and money planning in the hospital system, but some of the most vulnerable people in the community are in nursing homes and community-based living centers. The nation, she said, needs to make sure that there are incentives for ongoing partnerships and coalitions to help these other sources of care become more resilient. "We need more resilient communities, which means thinking about the social determinants of health," said DeSalvo. To her, that also points to the critical importance of increasing the resources available to public health as a means of getting resources into the community as quickly as possible.

Lessons from Hurricane Harvey

When Hurricane Harvey hit Houston in 2017, it dumped more than 50 inches of rain, or 19 trillion gallons of water, on the region, flooding the city, its bayous, creeks, and homes, explained Erin Erb. Nineteen tornados touched down to the north, west, and east of the city, all while sustained winds of 130 miles per hour buffeted Houston and its suburbs. In total, some 120,000 people needed to be rescued, and the storm produced a projected $200 billion in damage to the region. Erb noted that for the HCA health care family and coalition members, it was time to pull themselves up by their bootstraps and come together.

HCA's hospital in East Houston, the first to be evacuated, was built on a shipping channel and was destroyed by flooding. It will never reopen, said Erb. While the coalition was arranging to transfer patients, senior leadership in HCA's Gulf Coast Division was already at work on a reunification plan that would detail how to transfer those patients back into the city once the flooding had cleared. She noted that throughout the hurricane, business went on as emergency operations and preparedness drills became real life and plans were enacted for getting personnel and material resources to where they needed to be, with help from HCA's Nashville Emergency Operations Center, the Coalition's Catastrophic Medical Operations Center (CMOC), and the SouthEast Texas Regional Advisory Council (SETRAC).

One lesson from Hurricane Harvey was the importance of mobilizing teams to serve as reinforcements. "Never in my life would I have been able to imagine the sheer depth and breadth of what it would take to have this response," said Erb. By the time the hurricane had ended, 4 sister division emergency operations centers, 45 hospital command centers, CMOC, and 2 virtual transfer centers were involved in the response that involved 16 helicopters, 5 airplanes, 3 duck boats, 6 water tankers, 13 generators, and 82,000 gallons of diesel and gasoline. HCA marshaled out of its stores some 35,000 ready-to-eat meals, 6 days of linens, 2,000 20-pound bags of ice, 625 cots and air mattresses, 6,000 sandbags, and 120,000 bottles of water.

The response was tremendous, said Erb, but there were missed opportunities and lessons to learn. A serious challenge, for example, was coordinating air assets and landing zones throughout the storm, and communication among the coalition members was spotty at times, which led to duplication of efforts. After the flood waters receded, HCA's operating emergency departments were overrun by patients needing dialysis, highlighting the need to have a plan in place to address what was going to come after the storm. Erb explained that an additional lesson was to document everything that was happening as a means of informing postdisaster analysis. HCA learned, for example, that it needed to know the elevation of all its hospitals, the coordinates of its landing zones, and the location of each hospital in a specific flood plain. It also learned to invite its SETRAC coalition partners into its emergency operations councils so that they would be fully aligned for the next event. The final lesson, said Erb, was that the sun will rise again. In the end, she added, no patients, employees, or visitors were harmed.

Lessons from the Las Vegas Mass Shooting

Sunrise Hospital, said chief executive officer Todd Sklamberg, is the largest acute care hospital in Nevada, a regional tertiary center, and a level two trauma center. It is also among the closest hospitals to the Las Vegas Strip, so when he received a phone call at 10:20 PM on a Sunday evening that informed him of a mass casualty event, he went to the hospital not knowing what the community was facing. When he arrived at the emergency department, it was a scene like no other, he said. By the end of the event, Sunrise Hospital had seen 214 patients, plus another 30 that were treated and released before they could even be registered, and performed more than 83 surgeries within the first day. Of the patients who arrived, 92 had no identification, having lost their purses and wallets as they fled the fusillade of bullets.

Sklamberg explained that the hospital treated 124 gunshot victims, more than half of whom were brought to the hospital in private vehicles that had followed ambulances to the hospital or used mapping programs. One lesson learned from this experience was that it would be good to have a system that could direct those using cell phones to the proper hospital using geolocation. He recalled that his hospital received a call from one of the community hospitals who had a patient who had been shot in the head and needed a neurosurgeon. Unfortunately, there was no way to bring that patient to Sunrise Hospital in time.

At Sunrise, 16 deaths occurred. Ten were dead on arrival, four were beyond saving when they arrived, one was brain dead and had care withdrawn, and one died on the operating table. The hospital did not run out of blood thanks to receiving blood redirected from the non-trauma centers. At the time of the shooting, Sunrise's emergency department was already full, as were all 700 beds in the hospital, but by 6 AM, the hospital had discharged 180 patients. By 11 AM, emergency department operations had returned to normal.

As Sklamberg had mentioned, 92 patients came in without identification, and at one point more than 300 family members were in the hospital's auditorium looking for their loved ones. "We went through this the old-fashioned way," he said. "We sent staff up to the floors and took descriptions of the patients, had the family members give us descriptions, and sat in our board room and matched them." The two major identifiers, he noted, were tattoos and piercings. He added that the hospital was fortunate in that the VA and HCA both provided resources immediately. Within 24 hours, there were two VA mobile vans and grief counselors who could help both patients and staff members. Of the 3,000 Sunrise staff members, about 1,100 sought grief counseling and support, said Sklamberg.

Because the Las Vegas Strip is so close to the hospital, staff at the hospital have some experience dealing with surge capacity. It is not uncommon on New Year's Eve and New Year's Day for the emergency department to see more than 700 patients. The hospital's trauma surgeon and emergency medicine physician who were on duty at the time quickly decided how to set up a triage process that was able to parse patients into different areas of the hospital by the type and severity of their wounds. This enabled the specialists, as they arrived, to go directly to the appropriate area and triage their patients within the confines of one area, which reduced confusion and helped increase throughput to a degree that the hospital was able to continue accepting patients throughout the ordeal. One issue that arose was that the hospital ran out of mass casualty tags, so the trauma teams resorted to the old-fashioned procedure of writing vital signs on patients' foreheads with Sharpies.

An Ambulance Service Perspective

Forty-six years ago, with a degree in communications engineering, Richard Zuschlag founded Acadian Ambulance Service. Starting with two ambulances and eight medics who were Vietnam War veterans—before the era of licensed paramedics and emergency medical technicians—Acadian, the largest employee-owned ambulance service in the nation, now serves Louisiana, Mississippi, and Texas with 55 ambulances and 15 aircraft, and transports nearly 700,000 patients per year. Since 1971, Acadian has gone through almost 100 hurricanes and thousands of mass casualty vehicle incidents, train derailments, chemical plant incidents, and mass shootings.

From his perspective, the federal and state government responses to Hurricane Harvey had improved greatly since Hurricane Katrina. Zuschlag attributed that improvement to increased training, to the federal government getting everyone on the same page, and to the establishment of regional trauma-based emergency centers in Texas. During Hurricane Katrina, he said, many things went wrong, and both the governor of Louisiana and the President made mistakes. One reason for the difficulties encountered during Hurricane Katrina was the total failure of all forms of communication, including the entire wireless telephone system, the state police communication system, and the sheriff's communication system. In Louisiana, 18 parish 911 centers went dark when the storm came through, but because there was a statewide 911 call center, emergency services were able to continue to be dispatched. Nonetheless, miscommunication at all levels seemed to be the norm.

Given the challenges of finding safe routes into New Orleans in the aftermath of the storm, Zuschlag hired 30 petroleum helicopters out of Lafayette, Louisiana, to fly staff back and forth into New Orleans to help evacuate the hospitals there. At one hospital, his staff and the building engineer used flashlights to create a landing zone so the helicopters could land and evacuate critically ill infants. All told, his team was able to evacuate 80 babies from three major hospitals and take them to Baton Rouge.

Zuschlag recalled that when Laura Bush visited the communications center, he wrote a note to her husband, the President, explaining that no matter what was going on between him and the governor, the first disaster was the storm, the second was the levees being breached, and the third was the damage to the region's health care infrastructure and the need to evacuate not only hundreds of patients from seven hospitals, but hundreds of relatives who had come to the hospital, along with 85 dogs and cats. That note evidently got the president's attention for his chief of staff called at midnight, and Zuschlag was able to explain the situation, which was that the Louisiana National Guard, which normally would have provided help, was deployed to Iraq, leaving what he called the second and third string behind. "All of their barracks were flooded, and they did not have any vehicles, ammunition, guns, or radios," said Zuschlag. Hurricane Katrina struck on a Monday morning, and it was not until Friday that the U.S. Army arrived with satellite phones and support personnel that the hospitals were able to be evacuated.

Though everybody tried to do the right thing during Hurricane Katrina, the bottom line, said Zuschlag, was that too many things happened in the wrong way, largely because of a lack of communications capability. His concern going forward, with so much of the communication infrastructure moving to the Internet, is that a hacker could take down a big piece of the region's emergency communication systems. He credited FEMA's National Emergency Management Information System with doing a good job with training and developing protocols that have aligned public safety agencies, volunteer organizations, and public and private health care systems for disaster response. He was impressed during Hurricane Irma in 2017 with the way FEMA worked with American Medical Response (AMR) to enable companies such as his to ensure that paramedics are properly credentialed to work across state lines.

One thing Zuschlag has done in the years since Hurricane Katrina is work on interoperability of communication systems. He has also worked with the Louisiana Emergency Response Network to establish a system whereby every ambulance company in the state that is transporting a trauma patient can call the network's operations center to find out which hospital would provide the best care for a particular patient. In Texas, the regional advisory councils have done an excellent job organizing the response to mass casualty events and disasters, and their involvement during Hurricane Harvey was an immense help. One challenge that did arise was finding the right kind of watercraft to rescue patients. He noted that Texas and Louisiana are now working to properly credential members of the Cajun Navy, a volunteer organization that played a crucial role in rescue operations.

One thing Zuschlag noted was that every disaster is unique, and something always occurs that nobody had ever thought about before. One recommendation he made based on his experience with Hurricane Harvey was that the federal government could do a better job of having assets such as food and water, temporary shelters, security, and sanitation available for first responders. Other recommendations were to improve the mental health capabilities of the 911 system—mental health has become a bigger issue than most people realize, he said—and to develop a better system for organizing the Cajun Navy so they can use the Internet to register their boats and get apps on their phones so they can communicate with regional response centers.

DISCUSSION

Starting the discussion, Ricardo Martinez asked the panelists for the one thing they would do that would make the biggest difference in improving the response to a disaster. Cormier said he would institute better training programs at the coalition level to help with critical thinking skills for making decisions during a disaster. DeSalvo said she would boost funding for public health because it could be a huge first responder asset given its authority over safety, sanitation, air quality, and other important environmental challenges that arise following a disaster. "If we could have a mandatory funding stream for state and local public health, they would have more strength to not only surge, but be able to maintain partnerships and other work every day," said DeSalvo.

From his experience during Hurricane Harvey, Erb said she would work ahead of time to identify triggers that would be used to evacuate a health system before the situation becomes dire. Sklamberg's suggestion was to standardize cell phone power cords. His hospital now has power cords as part of its emergency stores of water and other supplies, but the multitude of connections means he must have a multitude of cords in stock. Zuschlag said he would require hospitals and nursing homes to have a comprehensive evacuation plan as a licensing requirement. In his experience, current evacuation plans call for moving patients locally, with no second or third choice if the disaster takes out those local facilities, too. As a result, there is often a great deal of confusion and delay in evacuating patients in regional-scale disasters.

John Hick asked Erb and DeSalvo to talk about the role of health care coalitions during the recovery phase and the opportunities for coalitions to engage in community planning for the recovery stage. Erb said that from the Gulf Coast perspective, HCA could have been better coordinated with the coalitions to understand what she and her team would be looking at after the storm and what the plan was for setting up makeshift triage tents, for example, in the emergency department parking lot or for serving dialysis patients who cannot get to their usual dialysis center. Although she and her team were able to work through these issues with CMOC, the real opportunity would have been to have those discussions before the storm hit.

Before answering Hick's question, DeSalvo raised an issue that arose during Hurricane Katrina regarding nursing home evacuation. It turned out that multiple nursing homes had contracted with the same ambulance company for evacuation capability, and when all the nursing homes requested evacuation simultaneously, the capacity to fulfill all the requests did not exist. Returning to the question at hand on coalitions, New Orleans did not have a formal coalition and depended instead on the health commissioner convening the community and having a good relationship with hospital leadership and emergency preparedness. These existing relationships mattered both during the storm and afterward in that they helped get the community health centers and ambulatory clinics involved to deflect surge visits to the emergency departments. Where this informal, human capital approach has not worked well was in dealing with people who are not in the hospitals or nursing facilities yet still receive care in their high-rise subsidized housing, so there needs to be better coordination with the home health agencies and other social services that have relationships with these individuals.

Having given up on the idea that electronic health record interoperability would ever happen, Arthur Kellerman of USUHS said his one wish was that every electronic health record vendor would at least populate a minimum essential database of 30 to 40 datapoints that every clinician could access in an emergency, then leave a record of their access to those data. To incentivize that, he would have Medicare, Medicaid, and private insurance companies require that every electronic health record has that feature in order be paid for medical services. DeSalvo agreed with his idea and said that eligibility to be a Medicare provider is an incentivizing mechanism that has not been leveraged regarding interoperability at this level. She also thought that Medicaid and commercial claims data could be used to identify the most vulnerable people and know where to send the ambulance or boat to get them.

Kellerman then asked the panelists, given their experience, what ASPR, HHS, and other federal agencies could do that would help their communities and regions to be better prepared for future events. Zuschlag's suggestion, which was less about preparedness and more about saving the federal government a great deal of money, was to let ambulance services use less expensive vehicles to transport people to walk-in clinics. In his area, for example, 30 percent of the 911 calls result in someone being transported to a walk-in clinic rather than the emergency department. Using a vehicle other than a fully equipped ambulance would be less expensive and produce a faster turnaround. Kellerman wondered if some of those patients could be treated at the scene, and Zuschlag said he would like to try a demonstration project in collaboration with a hospital to do just that.

Sklamberg noted that there are many efforts in Clark County and the rest of Nevada to better coordinate care and resources at the time of a disaster. What would help would be a system to provide real-time data from the scene and area hospitals, as well as a means of identifying patients, either through fingerprints or retinal scan. During the Las Vegas shooting incident, it took 6 hours to identify many of the patients brought to Sunrise Hospital, and none of the individuals who died had identification, making family notification difficult.

Erb noted that every HCA facility in Houston wanted armed guards to help deal with possible situations that might develop with the surge of patients and family members. HCA found itself in a fight with oil and gas companies to line up security services. "Next time, I would beat oil and gas to the punch," she said.

Ira Nemeth from the American College of Emergency Physicians and the University of Massachusetts Memorial Medical Center asked the panelist to comment on whether communication channels need to be formalized and structured or if a human capital, muscle memory, know-who-to-call approach is a sufficient or even better approach to maintaining communication during a disaster. Cormier replied with a story. During Hurricane Katrina, someone from New Orleans Charity Hospital called the emergency operations center requesting evacuation for its patients when the facility started flooding and losing power. Charity Hospital was told to move its patients to lower floors and be ready to be evacuated. Around the same time, the head of HCA's local division called a friend at the emergency operations center and was told that the situation was a mess and that HCA was on its own. HCA leadership decided to take charge of evacuating its Tulane Medical Center, while Charity Hospital struggled to evacuate its patients. "When you speak about formal communication, that formal communication has to be as honest as that muscle memory communication, and that is why we participate in national calls and regional calls, so we can listen and share information," said Cormier. "But until we can get the same type of information that we are getting from those one-on-one calls, I think we are going to struggle."

Zuschlag told another story in response to Nemeth's question. During Hurricane Katrina, one of his paramedics who was on one of his helicopters reported that five Blackhawk Helicopters were sitting on the ground at the Baton Rouge Airport and had not moved during the disaster. Zuschlag called his local manager and told him to contact the commander, which took some doing, but eventually he got on the phone. Zuschlag told him that the general in charge of the military response had put Zuschlag in charge of getting as many helicopters as possible down to New Orleans. Those five Blackhawks spent the next 3 days in New Orleans evacuating the sickest people out of the Superdome. Martinez added that social media, another form of informal communication, has become an important means for finding places for patients during emergencies.

An unidentified participant asked the panelists for any lessons they learned regarding security matters during a disaster. DeSalvo replied that one thing DoD brings to disaster response is that it does not have an agenda in the community and so it is not trying to vie for territory and be in charge, which she said is one of the ugly parts of disaster response. During Hurricane Katrina, the military's agenda was to do whatever was needed in terms of providing security and helping in whatever way it was needed. In her opinion, during a disaster when local government is overwhelmed, more consideration should be given to leveraging military assets because the military can come in and solve problems without trying to be the shiniest organization in the local community.

Michael Consuelos from The Hospital & Healthsystem Association of Pennsylvania asked DeSalvo for ideas on how he can get his emergency preparedness team to meet with his population team and talk about preparedness for future disasters and build more resilient communities. DeSalvo replied that Bloomberg Philanthropies has been funding work on how to build communities that are both climate resilient and socially resilient. This work has produced papers in the literature that can guide communities and health care systems. She also noted that there is an enormous amount of information that can guide the health care system and public health in efforts to build up resource-poor communities and individuals. Many private payers and Medicare are trying to understand how to use that information, and there are some "ready for primetime approaches" that are emerging to identify who is most at risk. For her, hospitals and health care systems should have an obligation to be part of the Medical Reserve Corps and volunteer to work with the local health department to go door to door and reach those outside of their walls during a disaster.

DeSalvo also expressed support and appreciation for the Public Health Service Commissioned Corps, which she said the nation underuses when running training exercises in the field. Martinez added that the public is another underused resource with regard to responders. He noted that when the Brussels bombing occurred in 2016, 100 civilians who had been trained by the European Red Cross went to the airport to help provide care for the considerable number of casualties. "We have an issue in this country where we rely on calling 911, which will always be overwhelmed in a disaster, so this is something we may want to look at as we go forward," said Martinez.

Ronald Stewart from the University of Texas Health in San Antonio remarked that there is no doubt that public health needs help and that public health will play a key role in the preparation and recovery phases. Unlike DeSalvo, he believes the first responders and acute care segment of the health system will be going into the community and bringing those who need care into the hospital. Given that, he believes that federal and state governments need to include hospitals and EMS as part of the command system that is involved in making decisions during a disaster. On the acute care side of the equation, Stewart said that hospitals and health care systems, no matter how prepared they are, are not self-sufficient. "We should shoot for an interdependent, diverse health care system that works together as a team," said Stewart. "Be as strong as possible but realize that on the scale of these events we're talking about, we are not going to be self-sufficient."

In Stewart's opinion, it is dangerous for a health care organization to think it is self-sufficient. Cormier responded that when he talks about being self-sufficient, he does not mean isolated. "We share resources and work with the federal government," said Cormier, and in his mind, being self-sufficient means that he has enough resources on hand so that government or other regional resources can be used at other sites.

Keeping with that theme, Sklamberg remarked that his institution could not have handled the Las Vegas shooting situation without assistance from the entire community, which meant sharing resources, sharing assets, and communicating where the best care for certain patients was. "If there had been discussion about doing this ourselves, we would not have survived," said Sklamberg, who again stressed the need for real-time coordination when disaster occurs.

Zuschlag said he was overwhelmed when some 500 ambulances from across the nation showed up and wanted to help in the aftermath of

Hurricane Katrina. With all communication systems down, it was challenging to get them credentialed and into the disaster areas. "It is amazing when something that big happens how the American people come forward and want to help," he said, and putting all of that help to work was definitely a team effort. He noted that although the number of deaths was tragic, the number of people saved is not discussed often.

Brendan Carr from ASPR asked if private-sector payers were part of the coordinating and planning effort. DeSalvo replied that payers did participate in the response during Hurricane Katrina. For example, the technology team from Blue Cross Blue Shield of Louisiana leaned in hard to create an electronic claims-based record for people in Louisiana to find information on the medications that people needed. That system, KatrinaHealth.org, was a significant piece of New Orleans's recovery planning to create a more resilient health care system. Erb said that payers were contacting HCA's service line vice presidents requesting information on where their patients were transferred, and she acknowledged that HCA did not do a great job of sharing that information in the moment as it was low on the priority list. HCA is looking at how to better handle those requests in the future.

Zuschlag said that during Hurricane Katrina, the Federal Aviation Administration (FAA) in Dallas called and asked his company to organize as many civilian helicopters as possible to help evacuate the hospitals. Though he responded immediately, 48 hours passed before he got authorization in writing so the company could be reimbursed eventually. He commented that the private health care community should not gouge the federal government, but unfortunately, many of his associates did. Perhaps because he was reasonable in the charges he submitted, FEMA and the FAA reimbursed him for services that went beyond the original contract, including flying food, guns, and ammunition to law enforcement and flying food to affected prisons. He noted, too, that FEMA and the State of Texas were far more organized during the 2017 hurricane season.

Mahshid Abir of the University of Michigan Medical School asked the panelists if they encountered any particular challenges during Hurricane Harvey in emergency care or inpatient care for pediatric patients. Martinez replied that his hospital had 30 children in his emergency department after a carbon monoxide event at a local school, pointing to the need to have the capabilities to treat multiple children during that kind of everyday emergency. In that case, his facility called on coalition members in the area to provide the necessary supplies and equipment. "A typical hospital does not have the expertise and the equipment to handle that kind of emergency," said Martinez. Sklamberg said that during the Las Vegas mass shooting, his hospital's pediatric subspecialists were helping care for adults, with one of its pediatric surgeons serving as a scrub nurse. "In times like that one, you use every resource you have," he said.

Commenting on DeSalvo's suggestion to use social determinants of health to identify those outside of the health care system's walls who need care during a disaster, Konduri said there are many efforts across the country to do just that. The challenge he sees is having the capacity in public health and social services to provide the care those individuals will need during a disaster. Phyllis Frosst from Squirus noted that ASPR's emPOWER initiative is working with Medicaid and all-payer records to look at the broader population in a community to identify who would be at risk during a disaster. One thing that has come from this work has been the good public- and private-sector engagement in the project, which can now produce lifesaving information in hours by leveraging Medicaid data.

DeSalvo commented that the nation is in the early days of developing financing models that support the clinical environment's efforts to identify and link people to social services resources, with Medicaid programs in Massachusetts, Minnesota, New York State, Oregon, and Rhode Island leading these efforts. In Kansas City, a pediatric hospital has created a one-stop shop for social services, and the American Academy of Pediatrics has a toolkit for identifying people in need and linking them to resources. Frosst then asked the panel for suggestions on what kind of information could help with preparations to reach those in the community who might need care in advance of a massive storm. Cormier suggested a layered map that would show patients the type of equipment they are using and what resources are responding to them, which could help reduce duplicated efforts.

Eric Epley from STRAC remarked that he has been involved in the responses to Hurricanes Katrina, Rita, and others, and the federal government's response to these disasters has improved greatly since Hurricane Katrina. He also supported the emPOWER project's work. He then noted that the San Antonio area is using its regional trauma and emergency health care system to support cardiac, stroke, and perinatal care as a means of not having to build independent systems. The area's newest effort is to use this infrastructure to help with mental health patients who have a history of needing help. This initiative is working with law enforcement to bring individuals straight to psychiatric hospitals, rather than to the emergency department, and using software to track social determinants of health and match patients with resources in the community. His question for the panelists was whether organizations dealing with social determinants of health, including EMS and acute care, should be involved in the health care coalitions.

Definitely, replied DeSalvo, who noted that San Diego and St. Louis, and probably others around the country, are taking a similar approach to San Antonio. She added that EMS and emergency departments often know much more about what is going on in someone's home than a primary care physician does. From her experience, it should be possible to build an information platform and navigation tools to identify who is going to bed hungry and who is dependent on electricity to power their medical equipment.

Gina Piazza from the Charlie Norwood VA Medical Center, the Medical College of Georgia of Augusta University, and the American College of Emergency Physicians' High-Threat Task Force asked how Sklamberg's hospital billed for the patients it treated and released without being registered or charted. Sklamberg said that Sunrise Hospital absorbed the cost of caring for those patients. In addition, the hospital has no plans to collect any co-pays or deductibles for a patient treated during that mass casualty event. That was not planned, he said, but it is a commitment that hospital leadership has made to the community.

To close the session, Kellerman recounted a story from the early days of Hurricane Katrina, when Admiral Thayer Cochran told the federal employees gathered in a warehouse in Baton Rouge who were struggling to coordinate the response to the storm that they should just worry about doing what is right, and to help the affected people as they would help their family and neighbors. If they had to break a rule to get something done, he would take the heat. Kellerman's question to the panel was how important is it to just do the right thing and not worry what the lawyer or claims adjuster thinks later. "This is something I think all of us who work in emergency care struggle with from time to time, where you have a lawyer perched on your shoulder whispering, 'Don't do that, you might get sued,' or 'Don't do that, you'll never get paid.'" DeSalvo replied that her husband said to her every day in the weeks after Katrina that she should just do the right thing until she got fired. Zuschlag said that as an employee-owned business, his company's attitude is that patient care and saving lives is job number one. "If the whole company went bankrupt because of it, so be it," he said.

Sklamberg echoed those comments and said he and his colleagues did not question themselves at any time during the response Las Vegas mass shooting. "It was always about the patient," he said, "and I am certain during the course of the event that there were HIPAA [Health Insurance Portability and Accountability Act] violations." He added, though, that health care systems also have to be responsible to families during these types of disasters. "Part of being a leader is providing hope and information, and even if it was not definitive information, giving folks updates on a regular basis is important," added Sklamberg.

Erb recalled how at one point during Hurricane Harvey, when multiple voices were arguing for different courses of action, the division president shut the door to the command center, muted the phone lines, and had a real discussion with the senior leadership team about needing to make decisions to do the right thing for its patients and for the health of the affected divisions. She noted that one big win for the leadership team was the decision to work with HCA's strategic communications group to push out "mission moments" that kept those in the field apprised of what was going on across the system.

Cormier said this is a struggle every day for clinicians in the emergency department who are scared to discharge a mental health patient or not prescribe an antibiotic. "But you see that drop during a disaster and you see competition drop in communities," said Cormier. The challenge, he said, is that the "right thing is not always this big glowing answer on the wall. Many times, it is a gray thing behind a cloud and somebody has to make that decision."

4

ASPR's New Vision for a Regional Health Response System

The last activity of the workshop's first day was a small-group discussion focusing on Assistant Secretary for Preparedness and Response Robert Kadlec's blog post on a regional approach to disaster preparedness and response (Kadlec, 2018). The mixture of public- and private-sector workshop participants at each of eight tables discussed four topics:

- What is the common understanding of the problem?
- How do we coordinate joint strategic activities?
- What are the most pressing needs regarding communication and coordination?
- How do we measure shared performance for shared accountability?

After an hour of deliberation, a rapporteur from each group reported back to the assembled workshop participants on the small-table discussions. In the first report out, Freda Gail Lyon from WellStar Health System said her table discussed shared performance and developed a list of terms that individual participants suggested needed common definitions: urgency, emergency, disaster, catastrophic, loss of governance, loss of mutual aid, loss of infrastructure, and loss of community function. Lyon explained that definitions for these terms are needed to develop metrics for measuring good management and the efficiency of the emergency management structure. Many of these safety and quality metrics, she explained, are already in use in health care settings. Individual participants also discussed how technology could enable the sharing of data and metrics to assess relative performance and ultimately move the field toward shared accountability.

John Hick's group divided the topic of shared performance into three categories: metrics, performance, and shared accountability. Hick explained that members of the group suggested that a preponderance of metrics are already available, such as regulatory metrics from the Joint Commission and CMS, and sets of metrics for engagement, process, exercises, and clinical care. Most of these metrics are useful before an emergency because measuring outcomes after an incident is more difficult.

From a performance standpoint, the group discussed completion of education and participation or engagement in coalition activities as potential metrics. Hick presented the idea that functional throughput and minimum supplies of certain pharmaceuticals could serve as useful metrics, too. The group explored the possibility of benchmarking against similar types of institutions, both regionally and nationally.

Regarding shared accountability, the group talked about the importance of regional councils and coalitions and the development of common policies and expectations formed through the development of relationships. Hick explained that shared accountability could be assessed based on after-action reports and the results of exercises, accounting for variations in size and capabilities among different institutions. The goal, said Hick, would be to create minimal expectations for facilities. Leadership accountability would also be important, the group noted. In the end, the group believed that success might involve tying preparedness activities to other time-sensitive emergencies, such as responding to stroke, trauma, or sepsis. In fact, this group suggested, there may be an opportunity to initiate pilot studies that would tie these existing time-sensitive emergency response systems into larger preparedness activities and serve as a means of driving investment in education systems for preparedness at multiple levels.

Laura Wooster from the American College of Emergency Physicians reported that her table discussed communication and coordination, particularly regarding whether health care coalitions have the resources, skills, and staff to engage in consistent and open communication across sectors for preparedness and response. Wooster noted the importance of looking for opportunities for daily use, such as communicating during the influenza season, using triage tags during smaller events, or using bed-tracking systems. One example discussed in the group was how the End Stage Renal Disease Network in Texas practices communication and coordination on a daily basis.

Wooster highlighted the group's discussion of the importance of including both emergency management and public health in these day-to-day communication and coordination activities. Participants also discussed the value of having plans for measured responses to specific types of events and to include contingencies for different types of infrastructure and communication systems failures so that there are backup communication plans.

Wooster's colleagues then explored how to build trust among organizations that typically compete with one another to ensure that everyone is working from the same guiding principles, shared vision, and common agenda. "Everyone needs to lay their cards on the table and take a leap of faith," she said, adding that a public announcement about what the coalitions are doing might be able to bring everybody together. Also important, she added, is making sure the same people are at the table consistently and to make sure those people are empowered to speak for their organizations. Repeated contact with one another is one way to build trust, she said.

Craig Vanderwagen's table also discussed the topic of communication and coordination and noted that everyone at the table immediately added the Wave 5.12 app to their smartphones,[1] giving them the capability to use a smartphone as a Motorola communication device among themselves. That action aside, the group spent much of its time discussing how to practice consistent and open communication for preparedness and response and how to build trust among organizations. Vanderwagen reported that whatever the solution, it has to be universal, bi- or multidirectional, transparent, and allow for dialogue between public and private entities.

Developing such a solution, he said, starts with understanding one's own role in the agencies that make up the public sector, which the group took to mean federal, state, and county governments. He noted that he would talk on day two about the conflicts that developed among different agencies within HHS as an example why defining roles is important. "If we are not clear about our roles in the public sector and in the private sector, it is very difficult to engender a dialogue between the public sector and the private sector that will move forward," said Vanderwagen.

Participants at this table opined that it is easier to work at the community level than at the hospital level given that hospital administrators have to spend most of their time dealing with day-to-day concerns and have little bandwidth to address larger issues. While it is important to help hospitals reach the necessary state of preparedness, it would be more effective to do that through community leaders who can then work with their hospital-based colleagues, Vanderwagen reported. He also noted the importance of understanding that leadership does not have to rest with one individual and that it can move among many individuals or organizations depending on the contextual reality. At the same time, it is essential to have a third-party broker, such as the coalition, a professional organization, the trauma surgeons, or other trusted authority, that can mediate disputes, both in planning for and responding to a large-scale disaster, he said.

In its discussions, the group considered the question of whether health

---

[1] See https://www.motorolasolutions.com/en_us/my-software/wave512.html#taboverview (accessed April 25, 2018).

care coalitions have the resources, skills, and staff to plan for and respond to a disaster, and participants suggested the answer was no overall. Yes, some organizations are well prepared, but others can barely function and those should be strengthened. One issue that came up during this group's discussions was that at least some private-sector organizations have no idea what the federal government brings to the table during a disaster. Participants suggested that ASPR could create a menu of options that would allow all coalitions, both strong and weak, to begin to understand what they can expect from the federal government.

In the end, said Vanderwagen, communication and coordination are the hardest pieces of disaster preparedness and response. "We can talk about mission, but if we cannot communicate and coordinate around that mission, it does not matter," he said. He noted that when he worked for the Indian Health Service, every staff member at every facility knew that their mission was to elevate the health status of American Indians and Alaskan Natives to the highest level possible. "We do not have that kind of clarity in this system because it isn't a system yet," said Vanderwagen, who added that building trust and communication channels will allow stakeholders to identify the common purpose of the coalitions. In his opinion, the ASPR blog post was a good attempt to get the coalitions headed in that direction.

Lewis Kaplan's group discussed strategic action. He noted that the federal government has recognized there is a gap that needs to be filled to the benefit of the American public. To that end, it may make sense to leverage existing infrastructure, such as the coalitions and National Trauma Programs identified by the American College of Surgeons' Committee on Trauma, he said. Marrying those requires a few important steps, including joining the trauma networks and public health regions so they are all working from the same plan, said Kaplan, who added that it was important to recognize that although trauma centers provide a model of bringing people together, not every patient should come to a trauma center. "The trauma centers can be leveraged to bring coalitions into more prominence so that patients are distributed and shared as appropriate during disaster responses," explained Kaplan.

Kaplan explained that while being a Disaster Center of Excellence is likely to be a money-making proposition in the future, it is not today, and it may be necessary to create incentives, such as a tax offset for unreimbursed care, to get large health centers to embrace the idea of being a Center of Excellence. Kaplan said that because this system needs to work exceptionally well, funding through ASPR's HPP that flows through public health departments may need to be changed because the current mechanism does not demand conformance to the HPP guidelines. The group discussed options to fund coalitions directly or have funds distributed in a competitive manner to those systems that do adhere to the HPP guidelines. Kaplan also highlighted that because law enforcement serves as the major transportation resource during a disaster, line-duty police officers might benefit from enhanced training to serve as competent first responders during a disaster.

Tener Veenema from the Johns Hopkins Bloomberg School of Public Health and School of Nursing then reported on her table's discussion about strategic initiatives. She said they suggested that there is a need for many distinct roles, strategies, skills, and expertise, depending on the event for which preparations are being made. However, she added, addressing workforce issues, such as education, training, composition, and sustainment, must be a major component of readiness in coalitions and a redesigned NDMS. So, too, is the development of leaders with crisis management skills, which she said the group members thought would be a key component of the strategic initiatives needed to build a bigger, better, smarter, and more coordinated disaster system.

Bruce Evans of the Upper Pine River Fire Protection District suggested that ASPR might follow the incident commander model developed by U.S. Forest Service and Urban Fire and Rescue. This model, he said, develops incident commanders through experiential learning that enables them to acquire the skills needed to serve as a member of an elite leadership team. He suggested these skills would give incident commanders the ability to go into any type of crisis situation, establish trust quickly, develop rapport, engender support, and execute decisions in times of increasing ambiguity.

Veenema's group also discussed designing hospitals as medical cities with residential living that enable a certain sector of the health care workforce to live on campus and be present at all times. Such an arrangement would reduce the need to move people in and out of the hospital during disasters, she said. The group explored using bloodmobiles and the blood supply system as part of surge capacity. This led to a discussion about using other health care resources that are not normally considered in surge capacity planning, but that could make a substantial contribution and affect a major shift in the way the nation funds and sustains preparedness.

Participants at this table ended their discussion with the recognition that the current health care system is optimized to the point where there is little wiggle room with regard to capacity and that when aligning NDMS with existing regional health care coalitions, additional demands are made of people who are doing their jobs every day. As a result, Veenema noted, it will be beneficial to mobilize and deploy health care provider teams from other parts of the country in the event of a large-scale or catastrophic event. The World Health Organization's emergency response team model might serve as an example for the United States as it develops regional, highly trained, event-specific strike teams that can be deployed rapidly at times of great disasters, she suggested.

Reporting on her group's discussions about a common agenda and a shared vision for the future, Gina Piazza said the public does not necessarily understand the gaps that exist in the nation's health care system and how they will affect the delivery of care during emergencies and disasters. The group also wrestled with whether medicine and professional societies are part of the public or private sector. She noted that participants believed that governments in general understand there is a problem with preparedness for large-scale disasters, but not necessarily at the appropriate level of detail. Piazza noted that some concerns were expressed that CMS might see the issues differently from ASPR, DHS, the Department of Transportation, or the National Highway Traffic Safety Administration, for example. Piazza went on to say that although hospitals may be aware there are challenges related to disaster preparedness, they tend to be focused on day-to-day operations and quarterly reports rather than large-scale and long-term planning.

With regard to a vision for future regional capabilities to respond more effectively and efficiently to the ever-expanding array of 21st-century health security threats, group participants expressed appreciation for ASPR's vision for the future as described in the blog entry. The group's discussion on regionalized health care pointed to the importance of defining regions for specific situations, Piazza reported. For example, she said, a region for burns might be different from regions for trauma depending on the specific capacities in those regions. Nonetheless, Piazza noted, a national emergency health and trauma system managed by regions that are networked to one another and connected to state governments would be the way to go. Then, in the event of a large-scale disaster that affected an entire region or two, they would be able to use the network to access additional resources. This pointed to the need for a resilient and adaptable regional system that can absorb a major hit and keep people alive, she said. As a final note, she said it might be important to develop standards of care for crisis situations that include provisions for relaxing some regulations so that providers are not so concerned about meeting HIPAA standards when trying to respond to a crisis situation during a large-scale disaster.

Ira Nemeth's group also discussed the common agenda. He reported that the group participants expressed that the public sector, at least at the federal level, has a good understanding of the gaps in the current system. He said the group did debate whether the federal government was willing to commit the necessary resources to address these gaps given other current priorities such as tackling the opioid epidemic. Regarding the private sector, the group talked about the fact that some private institutions have put a fair amount of work into preparedness and developing resources, while others are focused more on running their day-to-day businesses. Nemeth reported a common thread that larger health systems are likely to have a better understanding of the importance of preparedness, but that there was great variability across institutions depending on how many experiences they have had with large-scale disasters. Institutions along the Gulf Coast, for example, are likely to be more aware of the potential for large-scale disasters and therefore more likely to take preparedness seriously, he said.

When the group discussed a common vision for developing regional capacity to respond more effectively and efficiently, participants noted that the blog post referred to the trauma system frequently. Nemeth shared that the trauma system may be a good model in regions where it is already strong, but that preparedness is about more than trauma, and therefore, some individuals at his table were not very supportive of relying as heavily on the trauma system model. One alternative that might work in some regions would be to start by connecting community and smaller hospitals through the emergency medical system to larger tertiary facilities. Nemeth stressed that this would just be a start because it would also be important to include community partners such as home health care and dialysis centers in those coalitions.

One issue that was explored was that when talking about vision, there may be some misunderstanding that a specific institution would lead the coalition. That could lead to problems, said Hick, so it would be important to discuss what a coalition should look like and develop multiple examples of successful coalitions that different regions could draw from to meet the specific needs of their communities. Another challenge for establishing a coalition is getting buy-in from all of the necessary partners, and so it would be important to make the argument that joining a coalition can lead to cost sharing and reducing duplication of services, particularly for smaller institutions. Hick suggested that CMS regulations might also serve as incentives to join coalitions.

Nemeth reported that the group discussed how to include payers and insurance underwriters in developing a shared vision for preparedness. Underwriters, for example, might provide a break on insurance rates for facilities that belonged to regions that were more prepared to be resilient during a disaster, he said. Someone suggested that professional societies and local governments could be applying more pressure to hospitals and health care systems to join regional coalitions. The group even noted the possibility of conducting a public relations campaign that would help the public to understand how important preparedness is, which might convince consumers to apply pressure on their health care systems and local governments to take preparedness seriously and dedicate necessary resources.

Kellerman ended the session with a military phrase he learned when he became dean at USUHS: one team, one fight. "We may compete as hospital A, hospital B, and hospital C in a community for patients, and we may compete among the surgeons and emergency department doctors and the internists about who should get more reimbursement, but when a hurricane, an earthquake, or a Las Vegas mass shooting happens, it is one team, one fight," he said. "This discussion is all about how we assure that when the team comes together, it knows one another, trusts one another, and functions as a team and delivers as a team for the sake of our communities, our patients, and the national security of the country."

5

Looking to the Future

To start the second day of the workshop, Arthur Kellerman reminded participants that the workshop's overall focus is on large-scale disasters that far exceed the resources not only of a single hospital or local government, but even of a big national health system that might have the resources, but needs to network and coordinate with others in the response process. He also noted that an adequate response is built on existing and robust relationships, a workable structure, and confidence that organizations that might compete under normal circumstances will come together and cooperate when a community or entire region is faced with a large-scale disaster. For that reason, successful models more often resemble adaptive networks rather than hierarchical structures. Successful models, said Kellerman, combine the best attributes of individual institutions and communities and of local, state, and federal governments.

He then recounted a discussion he heard the previous day in one of the small groups about what determines a region with regard to forming a regional coalition. In Texas, for example, the regions were defined by looking at where patients self-referred and where physicians sent their patients for referrals. This reminded him of the old adage that the place to put sidewalks on a college campus is to look for the wear marks across the lawn. "Maybe that is a model we need to think about in the future for defining regions," said Kellerman.

There was a clear recognition from day one of the workshop of the need for regular interactions and drills, with no-notice drills being better than a dress rehearsal-type drill, Kellerman said. Similarly, near-misses and smaller scale events can provide insights of immeasurable value. There was also discussion that the players involved in responding to a mass casualty event may need to respond given the ongoing evolution of American health care from a primarily hospital-based model to a more disseminated, community-based model with outpatient surgery facilities, urgent care facilities, dialysis centers, and standalone emergency departments.

Kellerman noted that the National Academy of Medicine's concept of a learning health care system could be applied to the disaster response system. "I rarely use the word 'research' in disasters because the public does not understand how you can be doing research when you are trying to save lives and well-being," he said, "but concurrent evaluation could help us refine and incorporate and systematically get better, just as the joint trauma system did when the militaries were in Iraq and Afghanistan." The military's joint trauma system did not do research per se while fighting a war, but it refined, implemented, and adapted based on learnings in the field and became markedly better during the latter half of those conflicts than in the first half, he said. "We can do the same thing with disaster response in this country," said Kellerman.

One concept that came up, Kellerman summarized, was the need to relax some of the regulatory and legal standards that govern patient privacy, record sharing, and recordkeeping, for example. Another topic of discussion was whether the trauma system is the best model for coalitions, for while it is an appropriate model for some regions or situations, other mass casualty events could require a different model. A third theme Kellerman identified from the first day was that the federal government, including the VA and DoD, can contribute significant capabilities during a large-scale disaster that will enhance the community response.

The second morning's session featured three panels that addressed various aspects of thinking about the future of the nation's disaster response system. Open discussions followed each of the three panels. The first panel on best practices was moderated by Skip Skivington from Kaiser Permanente, and included Eileen Bulger from the University of Washington and the American College of Surgeons Committee on Trauma; James Jeng from the Mount Sinai Healthcare System and the American Burn Association; John Halamka from Beth Israel Deaconess Medical Center; and Gina Piazza from the Charlie Norwood VA Medical Center, the Medical College of Georgia of Augusta University, and the American College of Emergency Physicians' High-Threat Emergency Casualty Care Task Force.

The second panel, also moderated by Skivington, featured talks from individuals and organizations that are leading change across the field. The panelists were John Auerbach from the Trust for America's Health; Mitchell Katz from New York City Health and Hospitals; Ana McKee from the Joint Commission; and former Assistant Secretary for Preparedness and Response, Craig Vanderwagen currently of East West Protection.

The third panel, moderated by Jon Krohmer from the National Highway Traffic Safety Administration, included comments from individuals whose organizations are leading change at the local level. The panelists were Harold Engle from First Texas Cypress Fairbanks Medical Center Hospital; Lewis Kaplan from the University of Pennsylvania; Paul Kivela from the Napa Valley Emergency Medical Group and the American College of Emergency Physicians; J. Brent Myers from ESO Solutions and the National Association of EMS Physicians; and Paul Hinchey from Boulder (Colorado) Community Health.

CULTIVATING BEST PRACTICES

The American College of Surgeons Committee on Trauma, said Eileen Bulger, is a multispecialty organization of some 3,000 surgeons and surgical subspecialists. It sets the standards for what is required to become a trauma center and then verifies some 500 trauma centers per year. The organization also has a trauma system consultation program and a quality improvement program that engages more than 600 hospitals annually. Bulger explained that a trauma system represents the entire continuum of care for an injured patient, including injury prevention activities. From an acute care standpoint, trauma care begins with bystander intervention, she said, and toward that end, the committee has led the Stop the Bleed campaign that aims to provide civilians with basic bleeding control procedures. They define a trauma system as one that includes a trauma center, a communications system, and a close relationship with emergency medical services agencies. In their definition, care then extends into rehabilitation.

Trauma systems have also served as a model for emergency care for strokes, cardiac care, other time-sensitive medical conditions, and disaster preparedness. The keys to each of these systems are that they require timely and structured cooperation across multiple entities and that they are able to stand up on a daily basis, said Bulger. The trauma system, she said, is geared to deal with the everyday mass casualty events that occur in communities, but the communication and coordination that goes into responding to such events can provide the infrastructure needed to respond to a larger event. "When events get larger, the trauma system is able to surge in a way that can deal with a much larger group of patients, but still in a time-sensitive fashion," said Bulger. During a mass casualty event, she noted, the trauma system engages both the trauma centers and all of the acute care facilities in a region.

A high-functioning trauma system functions as an integrated learning health care system, said Bulger, who believes that such systems can serve as models for what the nation's disaster response systems should become. The trauma system is built on a public health approach, and good guidance is available on how to establish a trauma system (Committee on Trauma and Trauma System Evaluation and Planning Committee, 2008; HRSA, 2006). There are good data, she noted, that show trauma systems make a difference in terms of reducing mortality (Cudnik et al., 2009; MacKenzie et al., 2006). One recent study from Arkansas showed a 48 percent reduction in preventable deaths after implementation of a statewide trauma system (Maxson et al., 2017).

The Committee on Trauma advocates for an inclusive system in which all hospitals have to care for injured patients in a disaster setting, designed to get the right patients to the right place at the right time. "We want to match the patient to an appropriate center, and that means we have triage strategies, transfer strategies, and transfer agreements that make it clear how a patient should move through a trauma system," said Bulger. These agreements often cross health care systems, which requires establishing good relationships and communication strategies among those health care systems before a disaster strikes.

The committee also believes it is important to designate and distribute trauma centers based on the needs of the population. The goal, said Bulger, is regionalization rather than centralization, which requires defining a region based on patient flow and a network of hospitals that work together. She noted that states have taken varying approaches to creating a trauma system, with some having many trauma centers and others taking a more parsimonious approach (see Figure 5-1). Texas and Washington State, for example, take an inclusive approach that engages centers statewide to be involved in the system and have the appropriate training, infrastructure, and knowledge to move patients through the system. What this inclusivity accomplishes is that patients in rural areas will be able to reach a trauma center, be stabilized, and then transported if necessary to a higher level trauma center for further care. There are data, said Bulger, that correlate the structure of a trauma system to outcome and reveal marked variability in injury mortality by population based on system design (Brown et al., 2017).

In 2016, the National Academies of Sciences, Engineering, and Medicine issued a report calling for a national trauma care system that integrates military and civilian systems and optimizes system design (NASEM, 2016). The vision presented in this report, said Bulger, is to develop a system in which the civilian and military trauma systems share aims, infrastructure, system design, data, best practices, and personnel. She noted that the civilian system certainly benefits from the lessons the military learns in a conflict, but the military can maintain readiness by training with the civilian system when not engaged in a war. Together, the two systems working collaboratively would make the nation more resilient in the face of a disaster.

The Committee on Trauma has developed a list of the 10 minimum requirements for a trauma system that it is vetting with outside stakehold-

LOOKING TO THE FUTURE

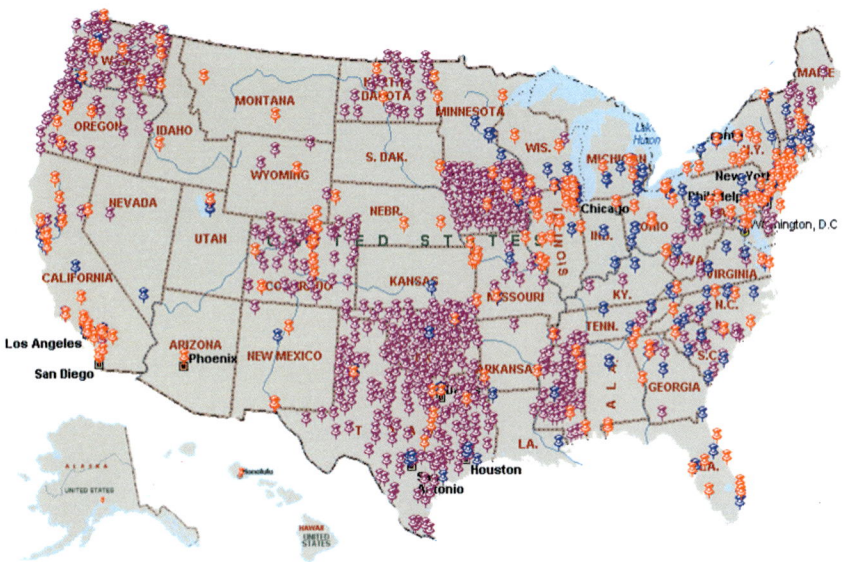

FIGURE 5-1 National distribution of trauma centers.
SOURCE: Presentation by Bulger, March 21, 2018.

ers (see Box 5-1). Bulger said that in the committee's view, the best strategy for trauma system development is to have minimum requirements set at the federal level that are then implemented at the regional and local levels. She noted that there is some literature on the impact of trauma systems on disaster preparedness, including a 2014 systematic review (Bachman et al., 2014). This review identified four key domains of a trauma system that affects disaster preparedness: communication, triage, transport, and training. STRAC's approach illustrates how an entire arm of disaster preparedness can be built from the trauma council structure, said Bulger.

In Bulger's opinion, trauma systems are the backbone for disaster planning because they offer a preexisting multidisciplinary governance structure, integration across health care systems, integration with EMS and air transport, established communication channels, and patient tracking strategies. In Washington State, for example, she can go online and see the status of every hospital in the state, get a picture of their bed status, and see changes in that status in real time.

In her final comments, Bulger listed the next steps for developing a solution to the nation's disaster preparedness deficiencies. The first step should be to bring to fruition the national trauma action plan called for in

> **BOX 5-1**
> **Committee on Trauma's Minimum Requirements for a Trauma System**
>
> - The trauma system should address the continuum of care and the needs of special populations.
> - Statutory authority for implementation should be in place (i.e., an agency with sufficient authority should take the leadership role).
> - A trauma advisory committee with broad stakeholder representation.
> - Creation, adoption, and regular update of a Trauma System Plan.
> - A process and criteria to designate trauma centers based on system need.
> - A funding mechanism for basic system infrastructure, including data collection/analysis.
> - Authority to collect and analyze injury surveillance data, at a minimum emergency medical services and trauma registry data from all acute care facilities.
> - Provisions for trauma system evaluation and performance improvement.
> - A trauma information management system with capacity to generate reports on system operations, quality metrics, and injury epidemiology.
> - Integration with military facilities, disaster, and mass casualty networks.
>
> SOURCE: Presentation by Bulger, March 21, 2018.

the National Academies report she mentioned earlier (NASEM, 2016). Doing so would require federal leadership to establish minimal trauma system standards and regional governance and implementation. There then needs to be support for trauma system development as part of disaster preparedness funding, something that Bulger said has been lacking since 2005, and optimized engagement between existing trauma systems and health care coalitions. She also explained that some mechanism is needed to facilitate the movement of licensed health care providers across state lines during a disaster, and trauma centers need to stand up and provide support to non-trauma centers, particularly during mass shooting events. She noted that the House of Representatives has passed the Mission Zero Act, which would provide funding to support the integration of military teams into civilian centers, and this bill is now awaiting approval by the Senate. Finally, the nation needs to support strategies for rapid assessment and debriefing after a major event to provide the lessons that inform improvement.

Addressing Strategic Bottlenecks

In his presentation, James Jeng pointed out that the burn care community represents one of several strategic bottlenecks in the national system response to mass casualties and other calamities. "We are a supporting actor, but the problem is there is not enough of us," he said. He reported that currently, there are only 1,800 burn beds and 250 active burn surgeons in the United States. "We cannot make more burn units or more burn doctors because it is economically not feasible in peace time, but in times of crisis, we need to be able to surge," said Jeng. The same can be said, he added, of orthopedic trauma, neurological trauma, radiation trauma, and other subspecialists where the economics do not enable the development of surge capacity. He called for the nation to develop a cogent strategy to deal with these bottlenecks, which he said are the weak links in the nation's trauma care system.

When discussing threats to public health and national security, burn care needs a place at the table to cultivate innovation and improve best practices, said Jeng. He noted that over the past 8 years, the American burn community has developed a rewarding partnership with ASPR as a result of Congress mandating that money be spent on preparing and amplifying the nation's burn response capabilities. Today, whenever there is a mass casualty event, or even when the President is speaking to a large crowd, ASPR and the Office of Emergency Management are in close contact with the American burn community. In fact, if ASPR and other federal agencies had not developed what Jeng called a symbiotic relationship with the U.S. burn care system, the nation would have no ability at all to amplify the burn care response in the event of a mass casualty. "It is this far-sighted thinking on the part of the U.S. government to help us come up with innovative ways to amplify to scale in the face of [a] 200, 2,000, or 20,000 mass burn casualty that allows me to sleep better at night," said Jeng. He added that in an explosion, roughly 25 percent of the trauma victims will have burns that will be life threatening without specialty burn care.

Jeng said that the burn care community recognizes its capacity is limited and that it will be on its own for the first 96 hours or so after a disaster. As a result, the burn community has developed its own mutually supportive, self-organizing network in which the members know each other's cell phone numbers and have an implicit understanding that they will only have each other to turn to for help in the first 3 to 4 days of a major crisis. He also noted that ASPR, through BARDA, has committed more than $500 million to private-sector research on burns, including the development of spray-on skin, a pineapple-based cream that melts away second-degree burn tissue before it becomes a full-thickness burn, and stem cells derived from adipose tissue. These technologies, which he predicted will be in the market soon, represent ways to amplify a fixed burn care infrastructure and give the burn care community hope that it can respond to a situation with 2,000 or even 20,000 burn victims.

One of Jeng's regrets is that the subspecialties that represent bottlenecks are siloed, so that each small community is doing its own lobbying for funds and developing its own protocols. These silos, he said, have the opposite effect of muscle memory when it comes to communication and coordination. "We need to start breaking down silos so that we are not islands of expertise but can be synergistic," said Jeng. Another key challenge he sees is getting these small communities to plan and prepare to foster on-demand scalability of strategic bottleneck supporting actors.

Regarding key elements that could be used to improve situational awareness of public- and private-sector capacity and capabilities to respond to disasters, Jeng said the burn community continually feeds information to the federal government about bed availability. Unfortunately, approximately 80 percent of the nation's burn beds are occupied on a typical day, and even moving some patients to normal beds would open only 500 to 600 burn beds in a crisis. He said the burn community is working with ASPR's Office of Emergency Management to feed detailed situational awareness data into ASPR's online geographical information system. As a final comment, he said the U.S. Army's Institute of Surgical Research has an irreplaceable role to play in the event of a mass casualty disaster that involves thousands of burn victims.

Information Technology in Times of Disaster

At 11:45 AM on November 2, 2002, a regional, integrated information network developed by a number of hospitals in the Boston area collapsed (Berinato, 2003; Halamka, 2008). "The hospitals of 2002 became the hospitals of 1902," said John Halamka. "We could not order anything, medications or labs. We could not view radiology studies. All of that automation that we depended on was gone." Within minutes of the system crashing, the 11 hospitals in the CareGroup network began running handwritten notes with lab results between hospital units, using telephones and every other non-Internet-based communication modality imaginable to keep their patients safe. The incident lasted for 3 days, and it provided a glimpse of what could happen if an electromagnetic pulse from a nuclear weapon took out all of the nation's electronics.

Halamka also recounted what happened in the aftermath of the Boston Marathon bombing in 2013, a regional incident that required care coordination, communication, and situational awareness. The mayor, however, ordered the region's cell phone network to be shut down because of the fear that the bombs were being set off by cell phone. "No cell phone communication was possible with any first responder," said Halamka, "and we had to move to alpha text paging, a 1970s technology, but a highly resilient technology in the absence of a cell phone network."

Another issue that arose in the bombing's aftermath concerned the security of victim and perpetrator records. Many people who came in with traumatic injuries needed pictures of their injuries for their medical records, and the surgeons used their smartphones to take pictures. "Imagine in those first few critical hours, potentially we had perpetrator and victim photography in the cloud, not where you would want it to be," said Halamka. "Since then, we have developed apps, secure photo repositories, and other things."

One year later, on the eve of Patriots' Day, the chief information officer of Children's Hospital received a message from the hacktivist group Anonymous informing him that they were going to take down the hospital's network as retribution for the hospital making one patient a ward of the hospital. The problem, said Halamka, was that Anonymous did not know the Internet address of Children's Hospital, so it launched an attack on an Internet subnet that included Harvard University, Massachusetts General, Brigham and Women's Hospital, Dana-Farber, Joslin, Beth Israel Deaconess, and Children's. At 2 AM on the eve of the marathon, every hospital in Boston lost the ability to communicate and coordinate care because the Internet of the entire region was flooded with a massive denial of service attack. At 2:10 AM, Halamka had every chief information officer on the phone. By 6 AM the network was back in operation thanks to Halamka writing a $60,000 check at 2:30 AM to a large technology company that offloaded the entire network.

Halamka and his colleagues learned several important lessons from these three incidents. One lesson was to use emerging technologies such as the cloud to increase resilience. Today, for example, seven petabytes of patient-identified information of all health care in Boston are stored in the Amazon cloud, a distributed, worldwide network with 50,000 employees. "In a mass casualty, who is going to be more resilient, a regional health care delivery system or Amazon?" asked Halamka. "Our answer is Amazon." Today, his health care system has contracts with a wide range of technology companies to provide cloud, mobile, interoperability, and machine-learning technologies. Recently, Halamka has been working with the Bill & Melinda Gates Foundation on a project in Africa that uses a blockchain mechanism to store data with great resiliency despite working in a region with poor infrastructure, unstable power, and governments that are not always honest. "You think about these kinds of emerging technologies that will give us as a health care system much more resiliency than any local actor," said Halamka.

Communication and cooperation has been a repeated theme during the workshop, and those are hallmarks of the network of health care system chief information officers in the Boston region. Halamka said they not only have each other on speed dial on their cell phones, but they plan together and drill together. They also work together on educating their workforces on downtime procedures, privacy imperatives, the latest malware and ransomware, and preparation for physical disasters. He believes the role of the private sector is increasingly important in the area of building cloud capacity and innovation and developing policies and protections so these emerging technologies can be used wisely. He pointed out that the Office for Civil Rights has been a great partner in figuring out how to keep data private and safe while moving it to these new, innovative platforms.

One of his jobs as an endowed chair of innovation at Harvard is to spread the lessons learned from incidents such as the three he discussed. In his talks around the country, he stresses that these new technologies are good enough and safe enough to serve as risk mitigators and tries to reassure health system administrators who are afraid of the cloud. His dream is that the early adopters of these technologies will join his efforts to convince everyone in health care that these technologies mitigate rather than increase risk.

High-Threat Incidents

Although a high-threat incident can describe any event that involves an ongoing physical threat to victims and responders, Piazza said she would be referring to violent criminal and terrorist-related incidents such as bombings, mass shootings, vehicle-borne attacks, and complex coordinated attacks. Such incidents disrupt public health, safety, and security. As an example, she described a 2018 event involving an active shooter. At the time, influenza cases were overwhelming emergency departments and all of her hospital's beds were full, as were those of every other hospital in the area. When she received the active shooter alert that indicated there was a strong possibility the shooter would enter her building with the intent to kill, her first thoughts were that her institution was not ready, and the region had insufficient capacity to handle mass casualties. "If we cannot do this well on a regular day, what can we expect on an exceptional day?" she asked. "With more than 50 years of attention to trauma and emergency care in a crisis, we remain a populace, a health care system, and a nation insufficiently prepared to deal with high-threat incidents."

In her opinion, the best opportunity to deal with such an event is to prevent it before it happens, but aside from that, the next best opportunity to intervene in the trauma chain of survival rests with empowered civilians. "The truest first responder is the closest able-bodied person to the significantly wounded casualty, and it would certainly behoove us as a nation to ensure that all able citizens know basic trauma first aid and how to respond in a high-threat incident," said Piazza, echoing Bulger's earlier recommendation.

High-threat incident response is complex and involves multiple disciplines acting under extreme stress, said Piazza. Unfortunately, the perpetrators of these events have shown the ability to learn the system's response tactics and evolve their threats to achieve maximum harm. As a result, the emergency response system must step up its efforts to learn from these events and adjust its tactics, medical interventions, transportation modalities, hospital preparations, first responder actions, and recovery elements to minimize mortality and also minimize morbidity to survivors, first responders, and the affected community and the nation that grieves along with the affected communities. One challenge to such a learning emergency response system is that there is no standardized methodology for gathering data around these events or a repository in which to place data for ongoing analysis.

To Piazza, DoD's work on reducing battlefield deaths represents a best practice for a learning health care system. By creating its joint trauma registry and populating it with data from nearly all casualties, the military was able to determine wounding patterns and improve both countermeasures and medical response protocols, generating best practice guidelines in the process. Once implemented, she said, DoD continued to gather data, study it, and refine those best practice guidelines. "This is superior to what we have been able to do thus far around high-threat incidents given our limitations as a disconnected response in emergency care environment in the United States," said Piazza. "Instead of learning by anecdote, news media reports, late and sometime redacted after-action reports and papers published with constraints, there might be a better way forward."

Looking at the public–private partnership around high-threat incidents reveals some barriers and possible levers, said Piazza. Mitigation remains challenging, she said, with breakdowns in communities and social relationships, insufficient mental health care services, and evolving criminal and terrorist threats. In her opinion, there is tremendous room for private-sector engagement in fostering community wellness and preparedness, though this would likely require new payment structures and aligning financial incentives around wellness, preparedness, and surge capacity. "The private sector and the general public could demand these changes from government," she said.

There is also room for a partnership with the public around preparedness. She cited a 1966 National Academies report that recommended that all U.S. school children past fifth grade should receive first aid training (NAS and NRC, 1966). "A prepared, educated, and empowered populous is a defense against high-threat incidents in the vein of see something, say something, and a force multiplier in response to high-threat incidents, i.e., do something," said Piazza. In her view, schools, community centers, public safety, hospitals, and businesses should offer such training, and insurance companies might consider lowering premiums for individuals who receive such training. Similarly, if true preparedness and surge capacity were part of quality measures or CMS conditions for participation, hospitals might be more apt to offer overtime pay for full engagement in full-scale exercises. As Piazza noted, readiness should not be an unfunded mandate and not intermittently funded by finite grants.

One best practice she would like to see adopted is to study these events and the response to them from the point of wounds through recovery. Such studies should become routine and include efforts to understand injury patterns and why some people live and others die. They should include studies of transportation methods and medical interventions by provider type. "We need to gather information as close to real time as possible and rapidly disseminate the lessons learned so that we are all better prepared for the next event," said Piazza. She also called for establishing a single entity with the authority to gather, store, and disseminate these data.

Building on the work of a number of colleagues, the FBI's victims services supply teams, and the National Transportation Safety Board (NTSB), Piazza's task force at the American College of Emergency Physicians suggests developing rapidly deployable multidisciplinary teams of subject-matter experts with the ability to gather discipline- and casualty-specific qualitative and quantitative data and develop best practices. Establishing such teams would likely require legislation and is another area, she said, where the public, a unified house of medicine, and public safety associations could bring their weight to bear to see such legislation passed.

In closing, Piazza asked how the nation can cultivate innovation and develop best practices for high-threat emergencies. She answered her own question by stating that the nation can develop an achievable strategic plan based on 50-plus years of excellent ideas, proper incentives, and objectives that are met and sustained. "We can employ a whole-community approach, as has been discussed over the past 2 days, and we can provide our researchers, educators, and innovators sufficient time, funding, authority, and opportunity for competition so as to foster success," she added. In addition, the nation can create a database and NTSB-style teams to facilitate learning and improvement and develop secure methodology for sharing lessons learned as rapidly as possible.

DISCUSSION

Lynne Bergero from the Joint Commission asked if the panelists had seen examples where training civilians to be prepared to respond has taken hold and been scaled. She explained that there are grassroots organizations in Chicago, where the Joint Commission is located, that have taken it upon themselves to undergo Red Cross training in first aid and that there are nascent community emergency response teams. She wondered, then, if it would be possible to ramp up this kind of grassroots training activity and connect it to other efforts such as the Stop the Bleed campaign.

Bulger replied that Stop the Bleed is focused on minimizing the most common cause of preventable death in these events. It consists of teaching very basic skills, such as how to pack a wound and hold pressure and how to apply a tourniquet. This campaign has two parts: public education that aims to train every citizen in the country, and having the necessary equipment in place similar to the way automatic external defibrillators are widely available. This program, which was launched by the Obama administration, has taken off under the leadership of the American College of Surgeons. Currently, some 150,000 people have been trained, and there are now 16,000 registered instructors. In her opinion, Stop the Bleed should be integrated into school systems and become a requirement for graduation, which is already the case in Washington State for cardiopulmonary resuscitation.

Piazza said the American College of Emergency Physicians is also a supporter of Stop the Bleed. She noted that the federal government has started the "Be the help until help arrives" campaign, and she is working on a follow-on project called First Responder On-Scene Training. Thomas Kirsch commented that the National Center for Disaster Medicine and Public Health is working with the Red Cross to roll out Stop the Bleed as part of its first aid training in the next few months. The Red Cross believes it can train 20 million people.

Piazza added that efforts like these to train law enforcement officers to provide initial medical care were resisted when she first started working with law enforcement, but that is no longer the case. Now, an increasing number of police officers are going into medic mode, as she called it, once the threat is neutralized, and then into transport mode to get patients to the hospital quickly. She said she would like to see more medical societies become involved in such efforts. "Imagine if every trauma surgeon and every emergency physician in this country taught one class to one group, how many people would be educated?" she said. "I truly believe that a nation prepared is going to be what allows us to have better outcomes in these events."

Bulger said that while training bystanders she realized this was not in the scope of practice for EMS. Her organization worked to change the scope of practice for EMS this year, and all EMS training now includes wound packing. For her, having as many partnerships as possible that are all saying the same thing will get the message out about the importance of being prepared to be a responder.

Kaplan asked Bulger to comment on the idea of using the trauma network as a framework for pandemic response. Bulger replied that her organization agrees with the criticism regarding using the trauma system as a backbone because not everything is trauma related. Having said that, the trauma system can be a model for how to set up relationships, bring a multidisciplinary group together that crosses health system boundaries, communicate regularly, and establish best practices on communication and moving patients through a system. To respond to pandemics, which are going to be much bigger events and involve more hospitals, the challenge is to build out that framework on a larger scale.

Monique Mansoura from the MITRE Corporation remarked that cyber security and information technology experts are not at the table with disaster preparedness leaders, and she asked Halamka for his ideas on how these cyber capabilities should be integrated into the broader preparedness domain. Halamka replied that cyber security is foundational to everything his health system does, noting that Harvard and its networks are attacked every 7 seconds, 24 hours per day, 7 days per week. These attacks are no longer from Massachusetts Institute of Technology students, which they used to be, but from organized crime, nation-sponsored cyber terrorists, and hacktivists who want to corrupt and damage data and steal identities. As a result, cyber security is at the table for every disaster planning activity his institution holds.

John Dreyzehner asked the panelists for ideas on how to deal with the silo problem that Jeng described. Jeng replied that while everyone complains about being constrained by limited budgets, breaking down silos is free and is a matter of collective will to not be beholden to turf wars. "Everybody in this room needs to espouse the concept [of] one team, one fight," he said. Bulger added that one symptom of having silos is the difficulty in combining registries with burn data, electronic health records, trauma quality improvement databases, autopsy data, and other sources of valuable information. The American College of Emergency Physicians has a big initiative to figure out how to assemble those data during a large-scale event in a way that is quick, comprehensive, and does not compromise law enforcement investigations to enable learning and improvement. She noted that there are no data showing how many people Stop the Bleed has saved, something that would be good to know.

To end the discussion period, Brendan Carr made a plea for the major professional societies to create a consortium and work together collaboratively to develop a list of next steps, a shared vision for what the federal government can do to further the nation's preparedness and response capabilities. "I hope I am not overstepping what can be said as a government official, but it would be helpful to hear a shared vision," said Carr.

LEADING CHANGE ACROSS THE FIELD

Skip Skivington opened the second panel in this session by noting that change is at the heart of what must happen to improve the nation's disaster preparedness and response, yet change is hard for both individuals and balkanized organizations. Nevertheless, he said, the four panelists all had experiences with fomenting change in their organizations. Craig Vanderwagen, the first panelist to speak, was the original Assistant Secretary for Preparedness and Response. He reminded participants that ASPR was stood up in 2006 after the passage of the Pandemic and All-Hazards Preparedness Act, and it occurred partially in response to the September 11, 2001, terrorist attacks on the World Trade Center and the Pentagon; the outbreak of severe acute respiratory syndrome in Canada that raised the specter of a looming pandemic; and the 2004 tsunami that devastated Indonesia, and Hurricanes Katrina and Rita. The Act, said Vanderwagen, effectively took into consideration ASPR's potential responsibilities in leading all federal assets in public health and medical disasters, including the formation of BARDA and the development of medical countermeasures for 15 threats identified by DHS.

The first problem he and his team faced in ASPR's early days was changing the culture of HHS, ASPR's parent agency. At the time, HHS was not an action-oriented organization, and its culture was dominated by two groups: the subject-matter experts, who generated the necessary evidence base but never had enough data to make a decision, and the contracting officials and lawyers, who in general believed, as Vanderwagen put it, that "you cannot get there from here." His approach to developing an action bias culture was to identify ASPR's mission and ensure that each person in the organization understood that mission, which was to save lives, reduce the burden of suffering, and speed recovery. "We managed and led around those objectives, and I say 'we' because it was a team effort," Vanderwagen explained.

ASPR's focus from the start was on facilitating local capacity building by incentivizing and nurturing the ability of local communities to own their outcomes in an effective way. He told how he grew up on a reservation in New Mexico, where his father's family had lived since the 1880s, and one lesson he learned living in a tribal community is that the community's survival and perpetuation of its culture and language are more important than the individual person. "We wanted to bring that kind of commitment to tribal survival to the communities we serve," said Vanderwagen. He also explained that the task of making BARDA an effective tool demanded that

ASPR engage with the private sector as a trusted and respected partner in a public health emergency countermeasure enterprise. This, too, required culture change, for the prevailing attitude in HHS at the time was one that undervalued and mistrusted the private sector.

Mitchell Katz explained that the New York City hospital system that he runs includes 11 public hospitals operating in the same areas as many private hospitals. As a result, he thinks a great deal about how private- and public-sector entities can work together. The best successes he has seen occur during mass casualty events, when EMS is in charge of transporting those who have been hurt to the hospitals best able to care for specific injuries.

The times when public- and private-sector entities do not work well, he said, is when one of the hospitals in a system goes down and patients have to be moved, along with their medical records, and clinicians need to work in a new hospital. The reason why that situation is so difficult, said Katz, is because nobody is in charge. "If I want 20 of my doctors to be able to cross the street and work at NYU [New York University], is that my responsibility to make that happen or NYU's responsibility?" asked Katz. "Who owns the medical records? Do I follow all of the discharge procedures as set in law for sending [medical personnel] in the midst of an emergency?" His answer to the last question was "presumably," but in none of the three cities where he has worked—New York, Los Angeles, and San Francisco—was there much in the way of planning or exercises that worked out how the public and private sectors worked together in that situation.

In New York City, the 11 hospitals Katz oversees work together among themselves, and he said he is sure the private systems do as well, but working independently of one another is not going to help if one hospital is knocked out during a crisis. He agreed with Jeng that collaboration is free, and he noted that money should not be a problem anyway given that the nation spends 18 percent of its gross domestic product on health care. "It is hard to say that money is the major reason we cannot respond in an emergency," said Katz.

One issue he would like to see addressed involved defining what an emergency is. To illustrate the problem, he recalled how New York City faced a shortage of pediatric Tamiflu, which commercial pharmacies ran out of in March 2018 during the influenza outbreak. He wanted to start providing this drug to his system's patients in the hospital and to waive the co-pay, reflecting the fact that many of the patients his system treats cannot afford the co-pay. His lawyers said he could not do that because the co-pay is contractual and is about decreasing usage. In retrospect, Katz suspects that if he had tried to argue that this was an emergency and that this contractual obligation should be waived, the lawyers would have pushed back that it was just the flu season and not a real emergency.

Katz noted that one of the most difficult things that happened during Hurricane Sandy had to do with credentialing and privileging doctors who could no longer work in their home hospitals. Normally, credentialing takes 2 months, he explained. Developing procedures that would shortcut that process during an emergency is something that should be worked out ahead of time. "We could decide what constitutes an emergency, and we could decide under these circumstances there are other ways of doing things," said Katz.

After noting that he agreed with an earlier comment that regions define themselves, he said that one lesson he has learned from every mass casualty event is the importance of having a well-defined leadership structure. Having such a structure is not about control, but rather is about making sure that things happen in an orderly manner. In closing, Katz said that as far as he knows, the three cities where he has worked do not have a leadership structure that would know how to respond to an emergency that included both the public and private sectors, although there is such a structure within each of those sectors.

Disaster preparedness and response, said Ana McKee, are critical to the mission of the Joint Commission, which she explained is the largest and oldest accreditor of health care in the United States. To the Joint Commission, emergency preparedness is about quality and patient safety, which puts it firmly in the organization's bailiwick. As an example of how the Joint Commission can be a resource in the national preparedness effort, McKee noted that when Ebola threatened U.S. hospitals, her organization knew that the hand hygiene rate of 40 to 50 percent across its accredited hospitals was insufficient to deal with that threat. The Joint Commission also knew from its data that the compliance rate regarding personal protective equipment was also low. If necessary, the Joint Commission could have sounded an alarm about this situation. "In many ways," said McKee, "we know what our organizations are capable of and ready to do."

The Joint Commission has established standards for emergency preparedness and was the first accrediting organization to require health care systems to plan and conduct drills for emergencies. McKee said these standards are excellent for being prepared to deal with a catastrophe within the walls of a hospital, and good if the disaster happens within a community, such as in a mass casualty bus accident. Where the Joint Commission, and the nation as a whole, needs to improve is in setting standards for regional disasters and for situations with massive infrastructure damage. In her opinion, a national leader with responsibility and accountability for pulling private, public, and government organizations together is needed to achieve the highest level of regional preparedness. Otherwise, she said, "the silos will kill us the next time something devastating happens."

McKee explained that the Joint Commission coordinates a lessons learned conference after every event that occurs in one of its accredited organizations. Those lessons are then put on its website and are available to everyone. The Joint Commission, McKee added, puts its learnings in publications and also goes on the road to train and educate its accredited organizations based on those learnings. For example, one of the lessons that came from leadership of Loma Linda University Health, which managed the response to the 2015 San Bernardino shootings, was the importance of managing an anxious workforce, something for which they had never trained. McKee said future standards will have to include something that will help leaders prepare for dealing with clinicians who are worried about their families during a disaster.

Along those lines, one lesson learned from Puerto Rico in the 2017 hurricane season was the need to take in families and their pets and to have the supplies on hand to have that capability. Other lessons the Joint Commission has learned from its organizations is that social media can play a vital role during an emergency, that a multidisciplinary approach to readiness is helpful, and that it is critical to shift from a hospital-centric approach to preparedness and response. In Puerto Rico, for example, a hospital learned how to use social media to find diesel fuel when the distribution system was destroyed.

In thinking about how to move forward, McKee said she is concerned about the capabilities of health system leadership. "What are the minimal competencies that a chief executive officer of a health system needs to have on preparedness?" she asked. "We do not speak about that." Nonetheless, that leadership will have to take charge of the organizational changes needed to improve preparedness. She is optimistic, though, that the standards and expectations that are being introduced to hospitals will produce the necessary change.

Commenting from his experience as the Boston health commissioner during 9/11, as the Massachusetts health commissioner during the H1N1 influenza outbreak, and at CDC when it was grappling with Ebola and Zika, John Auerbach offered five observations to the workshop. The first was that the public health sector is the key to preparedness and response. Public health is evolving, he said, to be more strategic and to bring together representatives of different sectors to deal with a wide variety of issues, but it should include emergency preparedness in its mission. Public health is well equipped to bring diverse sectors together, something that several speakers mentioned during the workshop.

Auerbach's second observation was that government resources are decreasing at a time when the number of large-scale emergencies is increasing. In 2017, for example, there were 16 emergencies that each cost more than $1 billion, yet hospital funding is 50 percent lower than it was several years ago and funding for public health is a third of what it once was, he noted.

Reauthorizing the Pandemic and All-Hazards Preparedness Act will be a key to increasing preparedness funding. Reauthorization should include a preapproved emergency preparedness fund for public health, similar to what the Federal Emergency Management Agency already has. The lack of such preapproved funds cost lives during both the Zika and Ebola outbreaks. "Lives were lost because we waited too long for Congress to pass the emergency funds that were needed to respond," Auerbach said.

His third observation was that it is important to think about the full spectrum of the public's health care needs beyond the walls of the hospital. Two areas that are frequently overlooked, he said, are behavioral health and long-term care. "From my experience in major emergencies, the most pressing issues that we experience are anxiety, misinformation, and the trauma that the public experiences, and we have to think about how to have the kind of services that include a basic service for anxious parents or family members as well as the ability to screen for people for whom that level of behavioral health issues is more serious," said Auerbach. During the hurricane season, for example, people in recovery who were in shelters were unable to continue to receive their substance abuse treatments, which became a problem. Regarding long-term care, he said the experience in recent regional disasters has been that the nation is not well prepared to evacuate nursing homes and long-term care facilities that are forced to close.

Auerbach's fourth point was that the focus is too often on the acute phase of an emergency, with little attention paid to the postemergency phase. He has heard from those who have gone through emergency situations that public attention often fades after the acute phase, and sometimes resources go away too, while people are still grappling with significant health and other problems such as trauma. His final observation is the importance of dealing with the social determinants of health, including poverty, the consequences of discrimination, social isolation, and challenges related to language skills. "What we have seen is if we do not pay attention to those, the most vulnerable populations often are the ones who have significant social and economic issues in their lives that make them less likely to know about what the response should be and less likely to have access to the services," said Auerbach.

Public health, he said, can play a key role in thinking about the specialized services that are needed to recognize the elevated risk and vulnerability of certain populations. In Boston after 9/11, for example, his office awarded contracts to train influential residents in subsidized housing on how to work with their fellow residents on pre-emergency planning. During the H1N1 outbreak, his office monitored who was getting vaccinated and noted that people who did not speak English were less likely to be vaccinated even though they were experiencing a higher level of risk of complication and death. His office was able to work with CDC to redirect funds to specialized community-based organizations that worked with those populations to get them vaccinated. "That required having a sensitivity and an ability to monitor whether those populations were in fact at an elevated risk and not seeking the services that were available to members of the public," said Auerbach.

DISCUSSION

Jeng started the discussion by coining a phrase—crisis standards for health care regulation—that he proposed using to get traction with both the Executive branch and Congress to reduce the regulatory burdens that lead to "we cannot do that" and the challenges of moving personnel across state lines during a regional disaster. Vanderwagen thanked Jeng for that suggestion and noted that the Pandemic and All-Hazards Preparedness Reauthorization Act offers an opportunity to develop legislative language that would be useful to this community. He encouraged workshop participants to think about possible legislative language that would highlight an area of need and enable the federal government to support local initiatives, rather than creating a structured regulatory environment.

Katz agreed with Jeng that the goal should be to have crisis standards, and he emphasized that the purpose was not to waive standards, but to have a separate set of standards that are appropriate for a declared emergency. He asked Vanderwagen if it would be possible to have that type of language inserted into legislation. Vanderwagen suggested developing a framework that would enable states to address legal, ethical, and medical issues relevant to local laws and make their own decisions, rather than forcing them into a strict regulatory structure. For example, he is working with Alaska to identify laws that are inhibitory and put ethical constraints on what may or may not be appropriate in a crisis environment.

Arthur Kellerman said that in addition to crisis standards, he believes there needs to be a focused approach to educating leaders who can lead in crisis, who can make decisions without perfect information, who can act without worrying about legal or regulatory consequences down the road, and who can work collaboratively and communicate with their employees and the public. Vanderwagen said one approach would be to put potential leaders in a situation where they have the opportunity to lead, to fail, and also to succeed.

Katz said that in his experience, physicians, nurses, and other clinical staff know what to do and focus on the right things, but the general bureaucracy tends not to allow it. He noted that there have been several times in his career where he decided that his staff would do something because it was right for the public and staff replied that they were worried they might get arrested or sued for doing it. His response was to say, "blame me." In fact, he has written letters ordering people to do things so that they could have proof that he ordered them to take a particular action during an emergency situation. "Sometimes, even if you are leading, that does not mean that everybody is willing to be certain that your endorsement will cover them if something bad were to happen," said Katz. Auerbach added that in his experience, getting support from an elected official is helpful in that situation. He also said that having high-level exercises and tabletop exercises that involve those who may called on to understand when those tough decisions need to be made can help lay the basis for a rapid decision later.

Harold Engle agreed with McKee that taking care of a hospital staff's family members was a major concern during Hurricane Harvey. Another issue that came up then was that many hospitals did not want to declare an emergency until the hurricane actually struck so that they would not have to pay overtime in case the hurricane skirted the region. In his opinion, there is an opportunity to mandate that health care professionals come in to work during an emergency with the caveat that their families would be cared for, noting that clinicians who are worried about their families would be less able to focus on taking care of their patients.

John Dreyzehner commented on the current practice for vaccination that is based on an opt-in framework where parents have to approve of their children receiving a flu shot, even during a pandemic situation. His concern is that this framework may make it hard to provide a medical countermeasure to the most vulnerable members of the nation's population and he wondered if an opt-out framework for medical countermeasures should be included in a crisis standard-of-care regulation.

Ronald Stewart raised the issue of recommending to the President that the nation needs a national trauma system to deal with the fact that 63,000 Americans, or 175 per day, die from violence every year. Brendan Carr, with the last comment of the discussion period, wondered if there should be a Public Health Emergency Medical Countermeasures Enterprise equivalent developed for the private-sector delivery system and private health care insurance sector.

LEADING CHANGE AT THE LOCAL LEVEL

In Houston, which has a dense population, hurricanes and flash floods are always a concern along with hazardous material spills, terrorism, and public health emergencies, said Engle. To prepare for such potential disruptions, he and his colleagues look at their institutions' hazard vulnerability assessments and their emergency operations plan, particularly after events occur. They also have 96-hour memoranda of operations in place to handle things internally before having to look for external resources. Engle explained that a hazard vulnerability assessment is done generally at the hospital level. It looks at a wide range of possible events that could occur and does a risk stratification based on how prepared the institution is to respond effectively to those events, how likely the event is to happen, and how many resources the institution can commit to the response. Its purpose, he said, is to prioritize the events for which an institution needs to prepare. He noted that he would like to see more regional involvement in these assessments so that the institutions in a region can work together more efficiently.

Engle's institution, First Texas Hospital, is rather small, but it has 20 satellite emergency departments around Houston that, in total, typically see 8,000 to 10,000 patients per month. During Hurricane Harvey, 18 of those 20 satellite facilities continued to operate at a volume 30 percent above normal. During the hurricane, many larger hospital systems closed their satellite emergency departments out of concern that they would have a patient who could not be transferred to a more appropriate facility. As a result, area EMS, which usually does not transport patients to freestanding emergency departments, started bringing patients to his system's facilities.

One tool that Eagle thought was particularly useful during Hurricane Harvey was SETRAC's EMResource, a Web-based tool that not only enables every health care facility in the region to post its capacity and how many patients it can take, but also make requests for needed supplies. Engle noted that SETRAC conducts tabletop exercises and drills that include home health, nursing homes, and dialysis centers. One thing that helped with Houston's response during the hurricane was that it never lost communication capabilities, but he is concerned that there is little redundancy in the system.

Regarding staffing, Engle said it is one thing to mandate that staff remain at a facility throughout an emergency, but there need to be provisions to have replacements after 36 hours. "You do not have replacements unless you call a disaster early enough and you do not call an emergency soon enough because your hospital does not want to pay for everyone to be at work," he said. During Hurricane Harvey, for example, his institution did not call the emergency early enough and some staff had to keep working for very long periods of time. Collaboration, as had already been mentioned, was critical for getting patients to safety during Hurricane Harvey. He, too, called for health care systems to have provisions to take care of family members of staff who have to work during an emergency.

One challenge with collaboration is when systems overcommit and overpromise and therefore are unable to collaborate during an emergency. Engle suggested that it might be useful to provide a summary reimbursement for unfunded care during a disaster, which he acknowledged would be expensive, but might encourage more collaboration. Other incentives might be to provide reimbursement based on a type of care scale or to provide tax breaks commensurate with the amount of unfunded care. Another incentive for local systems might be a disaster preparedness certification they could use in their branding. He also suggested that CMS could penalize systems that do not fulfill certain obligations and that the Joint Commission's standards could be tougher.

The Society of Critical Care Medicine has developed a disaster management plan designed to leverage member resources, said Kaplan. He noted that professional societies such as this one, which encompasses practitioners from several subspecialties, can be a valuable partner that regions and the federal government can call on during a large-scale disaster. During the crisis in Puerto Rico, the organization had to take a different approach because it could not get its members into the country. Instead, members gathered in Chicago, created an Amazon Smile account, and took in donated supplies. Those supplies were then collected, put onto pallets, and flown to Puerto Rico, along with one staff member, by a fruit-shipping business in Miami that was owned by a member's husband. That staff member oversaw distribution of the 80 or so pallets of donated supplies via all-terrain vehicles, off-road motorcycles, and on the backs of farm animals.

Kaplan's major complaint regarding Puerto Rico was that his organization did not have any direct contacts with the government, which limited the organization's potential impact. "There was a need for pediatric critical care nurses, and we have plenty of those, but we could not get a clear answer for how many were needed, where they should go, how long they would be deployed, what clearances they needed, and whether or not their license was appropriate because they came from many different states," he explained.

Shifting gears, Kaplan talked about the possibility of embedding a health care provider into a civilian tactical team as one way to build change from the ground level up. "We have heard about teaching law enforcement, but we have not heard about partnering with them in a robust way," he said. In his case, it took, as he put it, two cups of coffee, two donuts, and one croissant for his town's leadership to buy into this idea, but then it took 3 years for his university to okay the idea. The university, he explained, was concerned about workers' compensation, how his time would be valued, who would fill in for him when he was deployed with the team, how his participation translated into relatable relative value units, and whether his service with the team would count toward reappointment or promotion. Ultimately, the university decided this activity would qualify as an outreach activity of its trauma system. In Kaplan's mind, serving with the local tactical team was a way of demonstrating that the health care system was an actual partner that could help the team do its job better.

One benefit of this partnership was that he and others who participated in the program received training on how to use tourniquets and pack wounds that could be used on a daily basis when law enforcement deployed for non-routine but regular occurrences. Expanding this type of collaboration would be one way to guide change at the local level, said Kaplan, because it responds to a perceived community need while also increasing capacity for responding to disasters.

Kivela, speaking from his experience as an emergency physician in Napa, which has experienced two major fires, an earthquake, and mass shootings over the past several years, echoed Engle's comment that having to evacuate a hospital is a game-changing event. "No one is really ever prepared for that," he said. One lesson he and his colleagues learned during the evacuation of several hospitals during the recent wildfire catastrophe was that including a physician or a nurse on non-traditional transportation, such as a bus, facilitated the evacuation process. Kivela noted that while gaining experience is inevitable when responding to a disaster, he is not sure that learning is. To facilitate learning, he suggested that the federal government should create a home for emergency systems of care that would serve as a repository of lessons learned.

One big challenge he and his colleagues faced during the wildfires was that cell phone towers in the area burned down, making it impossible to communicate with his providers. In addition, hundreds of doctors and nurses lost their homes during the fires and have been unable to find housing. Many are therefore leaving the area, leaving it vulnerable to a subsequent disaster.

Myers, speaking as a member of the National Association of EMS Physicians, noted that he and the 650 board-certified members of the association look at the world from the community perspective in toward the hospital. Noting that multiple discussions over the course of the workshop talked about the need to create a public–private partnership with ASPR, he committed his organization and others to establishing a roundtable that would allow that type of partnership to be clearly channeled through physician specialty organizations. He also said his organization wants to work with the Joint Commission, which looks from the hospital outward, to change the mechanism by which drills are conducted and scored to include a region rather than just a single hospital. "We think that would be a wonderful mechanism to improve response across the community and enhance that public–private partnership," said Myers.

When it comes to crossing state lines, the EMS community developed the Recognition of EMS Personnel Licensure Interstate CompAct (REPLICA) program. REPLICA provides qualified EMS professionals licensed in their home state to a legal privilege to practice in another state. So far, 12 states have signed REPLICA into law, and Myers said he would welcome the federal government to encourage every state to adopt the compact. He also noted that EMS engages in public–private partnerships routinely, and he cited the public–private partnership to provide 911 responses in the District of Columbia as one of the best in the nation. Such partnerships, he said, can be amplified during disasters.

As a final note, he said it was important to consider the mental health of first responders during a disaster. Holders of an emergency medical technician credential, he said, are 1.4 times more likely to commit suicide than other members of the general public. "For our responders to be ready to enter into disaster, we have to support them in the regular time as well and then particularly after times of disaster," said Myers.

The final panelist, Paul Hinchey, pointed out that AMR, which has a contract with FEMA to provide resources in the event of a disaster declaration, is a notable example of a public–private partnership. Recently, he said, AMR merged with Air Medical Group holdings to become what might be the world's largest transportation agency, with multiple communication centers, including a redundant center it operates for FEMA during disasters. In addition to the resources it brings to a disaster, AMR becomes the structure framework that addresses silos and interoperability issues by facilitating the engagement and collection of local resources in advance of an event.

AMR, said Hinchey, believes strongly that local people should take care of local disasters, but it maintains a list of agencies from around the country that are willing to engage with and contribute resources in the event of a large-scale disaster that overwhelms local capacity. AMR has the ability, then, to reach out to those agencies and activate them under a coordinated structure that can deploy both during and after the acute response. In that instance, AMR becomes the link with FEMA that allows some degree of interoperability, he explained. He added that when the federal ambulance contract is activated, AMR serves as the facilitator to bring its resources and other EMS agencies to the response. AMR also tracks the movement of personnel and pharmaceuticals, including narcotics, across state lines in regional disasters in a manner that follows the intent of the laws governing those issues, said Hinchey.

DISCUSSION

Kaplan, referring to Engle's comment about wanting more regional involvement in conducting hazards vulnerability assessments, said that doing so in the context of a community partnership is essential because it then will include the perspective of everyone else who affects what happens at a given hospital. Involving the community would provide insights into how a nursing home, community center, first responders, and EMS view a hospital's preparedness, said Kaplan. Incorporating those insights into a hazards vulnerability assessment would improve preparedness because it would then reflect the interconnectivity of the institutions in a community, including those of state and federal partners. "If you do not embrace what they are bringing, you cannot prepare to interdigitate and interface with what they bring to you when you need it," said Kaplan.

Krohmer asked the panelists how they incorporate other local health resources into their plans for a coordinated response. Kivela, from his perspective on the EMS side, said he and his colleagues have established a community paramedicine program that looks at patients after they are discharged from the hospital. One thing he learned during the recent fires from patients who were evacuated from several hospitals is that many of just wanted to go home. He suggested that it might be worthwhile to think about using EMS to check on patients to make sure they are okay. He also noted that when moving vulnerable individuals out of evacuation centers, many were fearful they would lose the possessions they had with them, which points to the importance of meeting people at their preferred location during these types of events.

Myers commented that CMS will provide to any community the number of patients by zip code who are CMS beneficiaries and who have home medical devices that require power, oxygen, or other supplies. Having this information can help the regional public–private partnership to plan to have the required supplies and power sources or have transportation prearranged to meet the needs of those individuals. He also noted that during an event, EMS is likely to be the only source of care for individuals who are on a home health care plan, and regional plans need to account for that and not assume that those individuals will be able to receive care from their usual providers.

Kivela pointed out that freestanding emergency departments may be called on to provide surge capacity or pick up the slack if a hospital closes, but they are not eligible for reimbursement from CMS for Medicare or Medicaid beneficiaries. Engle confirmed that and noted that emergency departments attached to the CMS license of a hospital are able to accept Medicare, Medicaid, and Tricare patients. In Houston, he added, freestanding emergency departments are not required to participate in SETRAC and can decide whether or not to participate. SETRAC has a freestanding emergency department task force that is trying to determine what type of patients can best be served in those facilities so that EMS will be able to better determine where to take patients with specific needs during a disaster. During Hurricane Harvey, he added, those facilities were often the only choice for EMS to use.

Hinchey said it is difficult to use a facility that has never been used before and that has not participated in planning and drills. He recounted that when he was the medical director for Austin, Texas, the city engaged freestanding departments that were part of a provider network in all of the planning activities because they were deemed to be important receiving facilities. Those activities included helping the freestanding emergency departments be better prepared to receive and treat patients during a disaster who would normally fall outside of their usual scope of care. "If you are going to include them, the discussion needs to be held up front and they need to be part of the normal response," said Hinchey. "If they are not part of your normal response, it is always a challenge to throw something new on the table when everything else is coming unglued." Kaplan commented that the computing power of the U.S. National Laboratories should be harnessed to simulate what would happen if these facilities are or are not included in a response.

Kivela recalled there was a lack of communication among the hospitals, EMS, and the community during the fires in northern California. "People kept calling the hospitals," he said. "They did not know what to do or where to go." His suggestion was for health care providers to think about how they can use social media during a disaster to keep the community up to date on the situation and where resources are available for community members who need help. "The last thing you want is somebody going out in the flood trying to get somewhere and that place does not exist," he said.

Bruce Evans from the Upper Pine River Fire Protection District said that as a fire chief, his equipment is listed in a number of databases, include AMR's national ambulance strike team, the national interagency resource ordering and status system, a Web Emergency Operations Center (WebEOC) through the State of Colorado. During the Napa fires, FEMA made one call to AMR, which AMR made an immediate request for resources from his strike team, and a rig was on the road, headed to Napa within 12 hours. By contrast, when California declared a Stafford Act emergency for the Napa situation, that triggered an emergency management assistance compact request for 30 fire engines from Colorado. "That resource order came, and we finally had a truck that was able to get on the road 5 days later, by which time the damage was done in Napa," said Evans. In his experience from several incidents, the public–private partnership between FEMA and AMR is much more efficient than the government-to-government system. In his opinion, the fault lies with bureaucracy that gets in the way when two states are deciding who is going to pay for what, who has the approval authority, how many signatures are required, and who signs on the dotted line before somebody actually calls and says get moving.

Hinchey remarked that one problem with the FEMA–AMR system is that it conflicts with the disaster medical assistance team (DMAT) program that coordinates fire-based EMS. He said that many fire agencies, which have resources they want to contribute, are reluctant to sign onto the AMR system because they are worried about potential conflicts given that the two systems are not coordinated yet. Evans said that while he relies on

DMAT funds to supplement his usual funding sources, the AMR program reimburses him within 30 days, whereas the DMAT program takes months.

Kirsch asked the panelists what they see from the ground level as the ideal way to coordinate preparedness and response activities for health care systems. Engle replied that from his perspective, the regional advisory councils are an excellent mechanism for ensuring that hospitals and other institutions are getting the necessary resources. Myers agreed and added that the regional advisory committees are the most accepting group and by their very existence are an effective public–private partnership. Kaplan noted that regional advisory committees are effective, if a region has one. In the absence of one, a regional trauma center or regional coalition can serve as a coordinating body. Kivela added that regional coalitions are important for when a disaster encompasses many jurisdictions.

Melissa Harvey asked the panelists for their thoughts on what the optimal composition of a regional coalition would be and if it is possible for a coalition to be too big to be manageable. In Houston, said Engle, the coalition is still in its infancy and trying to figure out what kind of incentives it needs to get organizations to participate. "Without that or without some type of negative reinforcement, you may not get the participation because there [are] a lot of competing resources for health care dollars," he said. Kaplan remarked that a coalition needs a meaningful membership roster and a means of seeing itself as part of a larger framework.

Harvey then asked if a requirement for a coalition to fund a full-time position would help with right-sizing the coalitions. Engle said that might be helpful and having a full-time employee could lead to better allocation of resources at the local level. His hope, though, is that those joining coalitions would be doing so out of their desire to protect the public. Kaplan suggested that instead of funding a full-time employee, it might be required for each coalition to have at least two facilities that are in different tiers with regard to patient care.

Michael Consuelos noted that the different iterations of the Hospital Preparedness Program have driven degradations of capabilities in some areas and upgrades in others. As an example of the former, he said a hospital in suburban Philadelphia got funding during HPP's early years and was able to establish a decontamination unit capable of operating 24 hours per day in the event of a large-scale chemical or biological event. Today, however, that hospital reports that on a good day they can run the unit for 8 hours because both equipment and training have degraded over the year. His concern going forward is that as the coalitions get larger, funding and resources will be spread thinner and thinner. "I think it is important to understand at some point there are some critical pieces that we have to make sure are continuously funded and operational, otherwise, no matter how many people we have around the table, the critical infrastructure is not going to be there," said Consuelos. Another concern he has is making sure that the workforce is being trained adequately at the coalition level as the number of institutions and people participating grows.

William Wachter from North Central Baptist Hospital remarked that the most important part of belonging to a coalition is coming together to do something every day, whether that is holding classes and drills together, gathering data collectively, or participating in other activities. Tener Veenema added that the nation's 3.1 million registered nurses, as well as advanced-practice nurses, nurse midwives, nurse anesthetists, and nurse mental health professionals, will play a huge role in any response and therefore, nurses should be at the coalition table. She also noted that of the 3.1 million registered nurses, only 37,000 of them define themselves as public health nurses. "We have a potential shortage here for events where we need nurses outside of the hospital acute care health system," she said. Veenema also noted that most Federally Qualified Health Centers and stand-alone urgent care centers are understaffed, something that needs to be considered when thinking about these as potential resources for surge capacity.

Kivela remarked that there is a great deal of variance from one disaster to the next and one community to the next, which to him means that one solution is not going to fit all situations. "We need to do more research on this and debrief a little better after our disasters in the last several years and figure out what systems might work best and in which areas," he said.

Kellerman suggested that regional coalitions might look at large events such as festivals or major sporting events as opportunities to practice and exercise muscle memory. He also commented on Tennessee's approach to influenza vaccination, which is to let the community know that vaccination is going to happen at schools and that everyone can come and get vaccinated. To end the session, he then asked the panelists for one thing that ASPR could do to make their jobs easier. Engle replied that he would like more help making sure there is better collaboration among all the entities participating in a coalition. Kaplan said he would like ASPR to have a formal liaison structure for professional organizations. Myers said to clean up the Emergency Medical Treatment and Labor Act requirements during a disaster, and Hinchey requested a modification of existing legislative requirements regarding moving personnel, pharmaceuticals, and other supplies during a disaster. Kivela asked for better communication, collaboration, and learning during and from disasters.

6

Exploring Opportunities to Improve Private-Sector Investment in Capacity Building

The workshop's fifth session included two panels. The first examined regulatory barriers and facilitators for private-sector investment in capacity building and the second discussed financial barriers and facilitators. The panel on regulatory barriers and facilitators was moderated by Thomas Kirsch from USUHS and featured Alex Camacho-Vásconez from PAHO and Sean Griffin from the North American Electric Reliability Corporation (NERC). The second panel was moderated by Angela Brice-Smith from CMS and the four panelists were Andrew Baskin from Aetna; David Frankford from Rutgers Law School; Leslie Platt from the MITRE Corporation; and Lucy Savitz from Kaiser Permanente. An open discussion followed each panel.

EXAMINING REGULATORY BARRIERS AND FACILITATORS

Risk reduction is one of the main challenges that PAHO faces and one that it considers a key component of effective measures to approaching hospital preparedness in the Americas, said Alex Camacho-Vásconez. For several years, PAHO has been working on its Safe Hospitals Initiative and has created a hospital safety index that helps facilities assess their safety and avoid being a casualty of a disaster.[1] This index is based on structural, nonstructural, and functional factors, he explained, and the idea is for the index to provide a probability that the hospital would continue operating during

---

[1] See http://www.paho.org/disasters/index.php?option=com_content&view=article&id=964: safety-index&Itemid=912&lang=en (accessed April 27, 2018).

and after a disaster. A "C" classification, he said, means that the facility must take urgent steps to avoid failing during an emergency or disaster. A "B" classification means that intervention measures are needed in the short term, while an "A" classification denotes that the hospital will function in a disaster, though it should continue with preparedness activities. The most important reason to classify hospitals, said Camacho-Vásconez, is that it lets those who make decisions prioritize what they must do to have their facilities prepared for a disaster.

According to Camacho-Vásconez, a 2015 assessment of the Safe Hospital Initiative found that 71 percent of the 35 PAHO member nations had a national safe hospital program and 97 percent had created a database of evaluated hospitals using the safe hospital index. In addition, 63 percent of PAHO member countries had established formal, independent mechanisms for supervising hospital construction, and 80 percent had included safe hospital concepts in new health investment projects. The 2015 assessment also identified gaps, such as the fact that only 66 percent of the hospitals in PAHO member countries had up-to-date standards for the design of health facilities that were included in national building codes. The assessment also found that 20 percent of the hospitals had a "C" classification and 43 percent had a "B" classification.

To reflect the nature of hospitals in the Caribbean, PAHO has developed a set of tools to evaluate small and medium health facilities of the sort that are more common in Caribbean nations, explained Camacho-Vásconez. It has also started the Smart Hospital Initiative, which established a green checklist that will help hospitals be sustainable as well as safe as a means of reducing their carbon footprint and saving money. The first hospital to engage with this new initiative, the Georgetown hospital in St. Vincent, realized a significant drop in energy and water consumption in the year after revamping its facility to meet program requirements. Thanks to an investment of nearly $30 million from the United Kingdom, PAHO is intervening in 7 countries now and expects to intervene in at least 13 countries over the next 5 years.

Camacho-Vásconez noted that Mexico has used the hospital safety index to engage the private sector in efforts to improve the safety of its hospitals, and as a result, Mexico lost fewer than 50 beds during each of the past two earthquakes to hit the country, compared to the 5,000 beds it lost during the 1985 earthquake. He added that Mexico is taking the same approach with its schools and hotels and including them in its preparedness activities. Similarly, Ecuador's private sector has been actively involved in upgrading that country's hospitals. In a recent 7.8-magnitude earthquake to strike that country, the country's large hospitals were prepared, though the small- and medium-sized facilities were not.

NERC, explained Sean Griffin, is a nonprofit, international regulatory authority charged with ensuring the reliability of the bulk power system in Canada, the United States, and parts of Mexico. Originally, NERC was a voluntary group established in the 1960s after a variety of regional blackouts occurred in the United States and Canada. The bulk power industry decided to develop a set of voluntary guidelines and standards for the reliable operation of the three main power system interconnects. Following the widespread power outage that affected parts of the Northeastern and Midwestern United States and Ontario in August 2003, Congress decided to make the voluntary council a mandatory corporation and give it regulatory oversight. The Energy Policy Act of 2005 also empowered the U.S. president to work closely with Canada to harmonize standards across the two countries. Griffin noted that NERC's regulatory oversight is limited to bulk power generation and does not extend to the distribution system or retail-level sale of power. NERC does, however, maintain an information sharing and analysis center that covers all segments of the power industry and shares security and threat information relevant to the electric power sector.

NERC operates three committees—the Critical Infrastructure Protection Committee, the Operating Committee, and the Planning Committee—that develop the standards and guidelines meant to ensure a secure, reliable, and resilient bulk power system, said Griffin. The Critical Infrastructure Protection Committee, for example, develops cybersecurity standards, physical security standards, and emergency procedures for natural disasters, acts of terrorism, and pandemics. The committee process is driven by industry and is not by top-down federal edicts, an arrangement that Griffin said makes sense given that council members own and operate the relevant assets, which have a total value exceeding $1 trillion, and have a vested interest in making sure they are resilient in the face of disaster.

DISCUSSION

To begin the discussion, Kirsch asked the two panelists to describe any incentives and disincentives that have been effective at holding public and private entities accountable for the public good. One aspect of NERC, replied Griffin, is that it can fine violators $1 million per day, which provides a clear incentive for accountability and operating within the guidelines. Another incentive is the ability for companies to make the case for a rate hike to be fully effective at implementing standards and ensuring the electric grid is reliable. Florida Power and Light, the largest utility in Florida, made such a case after the 2004 and 2005 hurricane seasons. The resulting infusion of funds allowed the utility to invest in capacity that enabled it to reduce the average time to restore power following a hurricane from 15 days in 2005 to 2 to 3 days in 2017.

Kirsch then asked how rules, regulations, and rewards can be adapted or leveraged to improve public–private coordination, particularly for health systems that have many disparate missions and players. One challenge with applying the NERC model to health care, said Griffin, is that the power industry has one federal regulator, whereas the health care sector has many regulators. The NERC model could work, but it would require consolidating regulatory authority into one body, he said. In addition to regulatory standards, NERC also has reliability coordinators who act as the conductor of the power grid orchestra and ensure the grid operates as one large, integrated machine. Camacho-Vásconez added that the Safe Hospitals Initiative is not just about the safe hospital index, but it includes policies, programs, a budget, and other aspects that make a risk reduction program successful.

Kellerman asked Camacho-Vásconez if PAHO has evidence that Category A hospitals were more likely to survive disasters. Camacho-Vásconez replied that there are some data showing that the interventions have been successful in the British Virgin Islands, Dominica, Ecuador, and Mexico. He added that he has had personal experience that testifies to the effectiveness of this program. When he was director of the largest hospital in Ecuador some years ago, he made decisions on how to improve operations at the hospital based on the safe hospital index. During the subsequent H1N1 outbreak, his hospital successfully served as the national focal point for responding to the outbreak, something that would not have been possible without making those improvements.

John Hick pointed out that the health care system, unlike a utility operator, cannot go to a regulatory body to make a case for a rate hike when it needs funds to improve preparedness, for example. Griffin's suggestion was that if the health care system wanted to adopt a NERC-like model, it would have to work with federal and state regulators to reach a consensus on how the industry could recoup costs for such activities. He noted that a recent amendment to the Federal Power Act allows utilities to recoup the costs of responding to any national emergency declarations made by the Secretary of Energy.

Daniel Hanfling commented that a third model to consider is the one the VA used from 2007 to 2010 to engage in a comprehensive evaluation of its emergency management programs. The VA, modeling its approach after the PAHO program, developed an assessment tool with 71 capabilities that looked at both hospital-wide emergency management capabilities as well as network- and region-wide capabilities in the VA system. From that analysis of more than 150 VA medical centers, the VA developed a set of leading indicators for preparedness (Dobalian et al., 2013). He recommended considering the VA's approach to answering the question of whether its system was prepared to respond to a disaster.

EXAMINING FINANCIAL BARRIERS AND FACILITATORS

Health plans play a role in disasters, emergencies, and other large-scale disruptive events, said Baskin. In fact, he noted, every time a state or federal emergency is declared, it automatically triggers a response from Aetna that includes putting certain resources in play, depending on the specific event. When a disaster strikes, for example, Aetna opens its employee assistance program to any in the affected area that may need them, something Baskin suspects other insurers do as well. Within hours of Hurricane Harvey hitting Houston, Aetna had mined its database and had a list of every client who was on dialysis, had a home ventilator, required oxygen, was on a specialty infusion drug, or had other conditions that would require direct contact.

When Baskin and his colleagues have advance notice of an impending disaster, such as a hurricane, the company often will reach out proactively to all of its members who have a special need for care to arrange for early prescription refills, for example. Policy requirements, such as advance notice for hospitalizations, are put on hold during the event. "All of these things may not sound like the acute inpatient care that we have been talking about, but I can tell you that these are things that matter to the members out there," said Baskin. "They are calling us all the time because they want to make sure that their care is going to be covered and make sure someone is there to help them."

Baskin explained that Aetna has a cadre of nurses from around the country that it draws on to supplement local staff and ensure there is enough surge capacity to provide care during a disaster, including being available by phone for patients. All told, more than 1,000 case managers from his company alone can be mobilized online during an emergency to help patients receive the care they need or bring them to a place where they can receive needed care.

Frankford commented that the health care finance world has been well aware of the silos, lack of coordination, and the need for more community capacity for more than 40 years, dating back to the time when the Health Maintenance Organizations Act of 1973 was seen as a way to replicate health systems such as Kaiser Permanente and Geisinger Health System. What was learned from subsequent efforts, though perhaps not acted on, he said, was that capacity in organizations does not just spring into existence even if money is used as an incentive. He also noted that while consolidation in the health care sector may have its downsides, the upside is that coordination is easier to achieve.

In Frankford's view, emergency capacity in the health care system is a public good, something the public wants when needed, but hopes it never will. Today, however, the call to build surge capacity is an unfunded mandate. "We have to find a way to infuse money for this public good back into the system," said Frankford. Other wealthy countries, he said, have stable systems of financing emergency capacity that do not rely on appropriations from general revenues, in large part because they are usually an unstable source of funds. Rather, these countries use a dedicated source of funding, such as payroll taxes, trust funds, and surcharges on collected premiums. He cautioned, though, against trying to find some grand new design to support surge capacity because changing the payment system is like "trying to change the flow of the Mississippi River." Instead, he said, it is important to look for the tools at hand that can produce a dedicated revenue source for preparedness.

In the months after the September 11, 2001, attacks, Savitz received a grant under a bioterrorism preparedness program to create an atlas of nursing homes around the country. When Hurricane Katrina hit, those maps came in handy because the disaster management team had no idea where the nursing homes in the affected region were located. Savitz has overlaid maps showing locations of nursing homes, public health departments, major road networks, and funding regions for all 50 states and the District of Columbia. "There was absolutely no alignment or sensibility," said Savitz. "There was no strategic alignment of health."

Within 2 hours after officials received the maps, Savitz was asked to compile all of the data and drive to Washington, DC, where she worked at the Incident Command Center for the next 16 weeks to update the maps. At the time, these maps formed the basis of the only geographical information system (GIS) available for disaster planning, though today ASPR has one, too. She noted, though, that these maps are not complete because they are missing data on places of worship and other non-traditional providers of support. During the 2 years she worked in Biloxi, Mississippi, to help rebuild the primary health care system following Hurricane Katrina, she learned that places of worship play a key role in providing shelter and that pastoral nurses could serve as extra capacity during a disaster. The challenge, she said, is to consider all of the other community-based organizations that might be available during a disaster and to determine how to represent them in a GIS system so they can be called up during a disaster.

Turning to the subject of financing, Savitz said it took 7 months before Biloxi received any federal funding for recovery. In fact, the first funding it received came within 1 month from Qatar, which allowed the recovery team to begin the rebuilding process and get the federal quality community health center back in operation. That facility, she said, was the sole source of health care available in the community. Speaking from her perspective as a quality improvement professional, she said as money now flows into efforts to create big data resources and resilient communication channels, those efforts must be woven into the fabric of how care and services are delivered in a community. "That will allow it to be readily available, maintained, and practiced on an ongoing basis," said Savitz. "To have something standalone that we create and that we do not use is not going to be ready. It will not be maintained, and it will be a poor investment on our part."

Her final comment was that health care systems are now spending resources to serve the needs of populations, which takes them beyond the walls of their clinics and hospitals. In doing so, they are creating databases, and she suggested looking at ways to repurpose those resources for disaster preparedness by adding incentives, either through accreditation or other kinds of programming.

Platt explained that the issue "is not always about the money, but in this case the money is not at the table because we do not have the casualty insurers, the risk managers, or the investment community here." In particular, he said the investment community is "the big elephant that should be invited to the party." Based on 40 years of experience of working in various aspects of private capital, he believes there is a way to get the investment community involved to help fund a sustainable health system capable of responding to disasters. To him, the necessary surge capacity in terms of both physical and human resources, and the ability to make the health system resilient in the face of large-scale disasters, needs to be regarded as an essential cost of doing business. "It has to [be] part of the day job of every one of these certified participating facilities and organizational and institutional components of our ecosystem," said Platt.

With that framing, Platt said, preparedness, response, and resilience can become investable as part of the inherent cost of doing business. However, that message will only be driven home if preparedness becomes part of a health system's licensing and certification requirements. The Joint Commission, said Platt, has been moving in a helpful direction in that regard. In addition, the insurance industry has to change its thinking that disasters are relatively low-occurrence events given that they are becoming more frequent and have an outsized effect.

Platt's organization has been working on approaches to aligning the diverse interests of stakeholders in complex ecosystems such as health care. For example, MITRE has been working with the airline industry and the Federal Aviation Administration (FAA) to create an environment in which the airlines no longer compete on safety and instead pool information in a secure, redacted database that MITRE maintains. One requirement for creating such an ecosystem is for the stakeholder to accept that cooperation and coordination are inherent costs of doing business. In the case of health system preparedness, this means that stakeholders need to regard preparedness and resilience as a funding and financing obligation for the public good.

Platt said another essential aspect of achieving a committed financing stream for preparedness and resilience is to get the casualty insurers who provide general liability and director's and officer's insurance to the table. If a private health system, for example, is deficient in its preparedness or resilience and that puts the system at risk of failing its members, that deficiency can have immediate financial and regulatory consequences for the system and its executives, affecting liability insurance rates. "What I am saying is that if you start changing the framing for licensure, for certification, and for insurance risk classifications to include readiness, response, and resilience capabilities, then you get to a risk classification that actually matters," said Platt.

One approach to making preparedness and resilience investable is to declare them to be public goods with respect to national security and public health, Platt explained, which would bring together CMS, DHS, and DoD. Together, these three partners could align their efforts to fund performance breakthroughs that dramatically improve the capability of organizations to provide surge capacity, for example. FEMA already uses provisions in the Economy Act to encourage federal agencies to collaborate and provide funding for activities related to its mission.

The bottom line, he said, is that the financial community would rather invest in building and rebuilding American infrastructure. The key is giving the financial community investable instruments that are fairly harmonized and standardized, such as bonds that are rated by bond-rating organizations. "If you line up the funding streams so that [they are not just nice things] to have, but [they are] actually investable, we can actually change the delta on this so that the next time there is a meeting like this, we can actually have the financial folks excited to be looking at a brand-new deal flow where they can make fees on investment," said Platt. His advice in closing was to follow the money. "Follow the tried and true mechanisms for aligning the capital to make this a national priority as part of our national security," he said.

DISCUSSION

Brice-Smith asked if Baskin had any ideas from the perspective of a health insurer on how to better deal with some of the financial issues related to threats and disasters. Baskin replied that the cost of emergency preparedness, response, and resiliency is more than just a medical cost issue. Today, health plans are already financing the medical costs that occur, and in that respect, he has some concerns going forward as the health care financing system starts moving more risk onto providers than it has in the past. "When we talk about alternative payment models where we are transferring risk, we are transferring the risk for disasters as well, and we are transferring these medical costs, which could surge and potentially harm a provider organization in one of those arrangements financially," said Baskin. Although some protections are available to providers, such as stop-loss insurance, for a disaster that raises cost of care by 25 percent, for example, even a 5 percent increase over the contracted amount could be devastating to an organization that operates on a 1 to 2 percent margin, he explained. "I do not know that we have figured out the mechanism to protect against that," said Baskin.

Savitz agreed with Baskin and said that in the drive to purchase by moving into a risk-based environment, nobody has really thought through the implications of assuming that risk on the downside. Frankford noted that the U.S. health care financing system has been built around shifting costs and risks to someone else. Although the Patient Protection and Affordable Care Act ameliorated that to some extent, shifting costs and risks are still basic incentives in the system. "This raises the classic free rider problem, where everybody looks to everybody else to make the necessary investment, and no single entity alone has the incentive to make that investment because they thereby confer benefit on their competitors," said Frankford. "What we need here is collaboration across institutions." He agreed with Platt that part of the answer is to bring in private capital, but that approach has limitations in a system with such massive fragmentation as is the case in health care.

Before opening the discussion to the workshop participants, Brice-Smith noted that the private sector's response during Hurricane Harvey was tremendous in terms of pitching in to provide supplies, medicines, and clothing. Southwest Airlines, she said, flew in to evacuate people, which surprised her. In her mind, that type of response given the circumstances requires a great deal of coordination, which she found impressive.

Savitz commented that health care has changed dramatically since Hurricane Katrina in that more care is provided in the community as hospital stays have been shortened and unnecessary emergency departments visits and hospitalizations have been reduced. Much of that care is provided by informal caregivers who are funded in many instances by Medicaid or Medicare. She wondered if this workforce can be trained to provide what she called preparedness in place. She also wondered if CMS's Healthy Communities Initiative could be leveraged to improve preparedness.

Ira Nemeth asked Baskin if there was some way to involve the insurance industry in better coordinating activities during and after a disaster. Baskin replied that he has wondered if there was some way the insurance industry could develop a standardized way to communicate with the broader provider community, share information better, harmonize policy liberalizations for disaster situations, and provide consolidated information about resources that are available in the community, such as telemedicine services for behavioral health. "As a group of health plans, could we somehow share those resources and make them available across lines of health insurers?" he asked. "It certainly makes some sense to me that what we are doing can be leveraged and more efficiently provided and be more uniformly communicated."

Nemeth also asked if there was a way to fund preparedness through some sort of universal levy, in the way that universal telephone service in the United States was funded by a small tax added to telephone bills. Platt agreed that something different needs to be tried. In his mind, forming consortia that would raise funds together make sense. His company, for example, offers a catastrophe bond, underwritten by Moody's, to protect against the risk associated with hurricanes and other catastrophes that might be appealing to a consortium of health systems. He sees now why some of the cost of buying that kind of bond could not be part of the premium paid by CMS or other parts of the federal government. The point, he said, is that there are incentives for the financial alignment of performance improvements using consortia as the vehicle for harmonizing stakeholder values. "The tools are there, the capital is there, and the risk management, insurance, and reinsurance vehicles are there," said Platt. "They are not new."

Savitz countered that health care is not perceived as a social good in this country and that a use tax tends to be a regressive tax. Baskin said he agrees with the use of the carrot, but he would like to add the stick. He noted that health plans excel at creating networks, which involves coordinating among providers of diverse types who are responsible for providing health care for a population. Moreover, providing health care to a population is an obligation to have capacity for national emergencies, he noted. Given that argument, he said he could see state insurance regulators requiring investments in preparedness and resilience. Frankford seconded this idea of using network adequacy as a stick, particularly with regard to coordinating the components of a network, which he believes is the missing piece regarding preparedness and response.

Kaplan, returning to Brice-Smith's remark about the scale of the private sector's voluntary participation in the response to Hurricane Harvey, said the American spirit of volunteerism is something the health care system has not tapped or leveraged yet. The problem, he said, is that no vehicle is available to align the companies that jump in to provide support during disasters. He wondered if there might be a way of enlisting the private sector as formal partners and develop a vehicle so they can help fund some of these initiatives. "They already play in the same sandbox when needed, so why not have them as our proactive partners rather than our retroactive supporters?" asked Kaplan.

Laura Wooster, from her perspective as an emergency care physician, asked if there was a way to use the community benefit tax deduction as an incentive for health systems to invest in preparedness and other emergency care activities. Baskin replied that the current community benefit requirement is voluntary to some extent, which he acknowledged is oxymoronic. He added that the type of capacity this workshop has discussed is currently counted as a community benefit, but coercive legislation is needed to force more capacity building to occur. The limitation, though, is that the community benefit deduction is only useful to the not-for-profit sector.

Helen Burstin from the Council of Medical Specialty Societies wondered if it would be easier to model what an approach could look like at the community or coalition level by considering an influenza outbreak, which fits more easily into a health care model, rather than a big natural disaster to see if there are some structural or process metrics that could be used for accountability. She also wondered if it would be possible to model the potential savings that could be gained by having coalitions that work together to mitigate poor outcomes and long-term financial effects of an influenza pandemic, for example. Baskin replied that influenza is essentially a planned, annual event that lends itself to a different type of approach. Nevertheless, work has been done with influenza that could be applied to a one-off disaster. For example, insurance companies have expanded their concept of networks to include pharmacies and community health centers as places that could provide flu vaccines and developed new mechanisms that made it simple for these non-network providers to bill and be reimbursed. In the same vein, he wondered if it would be possible for health plans to cooperate and collaborate on their networks so that the network can be as big as it needs to be to fit one of these large-scale disasters. "This makes total sense to me, but I am just not sure of the form to accomplish it," said Baskin.

Platt said that MITRE has been developing a national strategy for accelerating and expanding private co-investment in biomedical innovations that can deliver breakthrough improvements to standards of clinical care for infectious diseases, including influenza, as well as many other large-burden exposures that are heavy cost centers for CMS. "In that context, we have done extensive work to demonstrate that CMS has the authority today to be not just the world's largest insurance payer for current standard of care, but also the world's largest innovation buyer by promoting target product profiles for universal flu vaccines and other breakthroughs, such as a replacement for end-stage renal dialysis," he said. In his opinion, the same authority could be applied using the logic that the public good is a reduction in the likelihood of CMS exposure to massive losses from other large-scale disasters.

Consuelos remarked that securitizing private investments around population was a hot topic several years ago, but the results so far have been less than expected. He also noted that his emergency preparedness team had tried to speak to leadership about the role preparedness can play in reducing risk and associated insurance premiums, but there seems to be a complexity to this issue that chief financial officers cannot seem to grasp. Given that, he asked the panelists for their suggestions on how to make the argument for preparedness in a way that chief financial officers can understand and place a higher priority on these activities. Savitz replied that when she was a financial planner for a health system, preparedness was not the number one thing on her list of concerns. "I hate to be so crass about it, but people are responding in crisis management mode to what is in front of them right now." She noted that she recently moved to Portland, Oregon, which has a large number of bridges that are susceptible to earthquake damage, which would isolate communities. As a result, preparedness for an earthquake is high on everyone's priority list in Portland.

Platt said one reason securitization has not yet been easy to do in the health care environment, as well as in other areas of social impact, is the lack of enough scale and harmonization to create standard vehicles that minimize the amount of work underwriters have to perform. "Wall Street is looking to put a lot of capital into this arena, but they are looking for harmonization mechanisms," he said. Platt then noted that unfunded mandates and unfunded, unfinanceable vehicles are not the answer to financing preparedness. "The whole system has to start moving incrementally to build the infrastructure, so we can start to improve the way in which it performs," said Platt. "What I am suggesting is that there is a path by which to build essentiality as part of the cost of doing business into the role of emergency surge capacity medicine." In his opinion, there are steps that would align the interests of the private sector and capital markets and deliver funding and financing that makes this sustainable.

Baskin said the real question that needs to be asked is about the role of a health plan in the future in terms of broader emergency preparedness and financing. "What services do we bring to the table that could be coordinated with other services to make this an additive type thing?" he asked. "I think that is a big issue, and I do not think it is well defined." One opportunity he sees is for health plans as a sector to talk among themselves and then talk with health system and federal leadership about what health plans can offer in terms of communication capabilities, data analysis, or dollars that could enhance the ability of the provider community to do what its job when the times come. An unidentified participant noted that the insurance industry was a terrific partner during the H1N1 influenza pandemic, but the health system does not really understand how to engage the insurance sector on a more sustained basis to work collaboratively on these situations.

Bergero noted there is a huge lever with regard to increasing surge capacity and resilience that has not been pulled at the national level, and that is the potential role of home health care. This type of care, which includes skilled nursing care, respiratory therapists, oxygen providers, and durable medical equipment providers, is the Joint Commission's second largest accreditation program after hospitals, she said, and she suggested that a national convening organization, whether it is ASPR or some other body, should bring home health care to the table to talk about strategic issues such as surge capacity. While home health used to consist of small organizations spread throughout a community, it has become much better organized in recent years and now major national associations exist, such as the National Association of Home Care and Hospice.

Piazza commented that her community was about to have a large sporting event in its area and while the event would be hiring public safety and medical staff to be at the event, local hospitals have to increase staff, without compensation, to be prepared for a disaster at the event. She wondered if state and local government or the private sector should be responsible for covering those added costs. Platt recommended that local hospitals need to speak up during the planning for such events so that their costs are considered part of the event sponsor's expenses.

The final question of the session came from Kellerman. He asked Baskin if there was some mechanism that could be used to reimburse pharmacies if they were given the authority to dispense Tamiflu during an influenza outbreak without first requiring individuals to go to the hospital or to see their physician. Baskin replied that health plans reimburse for services allowed under the scope of practice associated with their license. If states were to include that ability in their scope of practice, there would not be a barrier to reimbursement from a health plan point of view, he said.

7

Final Thoughts

Assistant Secretary for Preparedness and Response Robert Kadlec concluded the workshop with a keynote presentation that focused on the notion of a public–private partnership being critical to many of the programs at HHS. Public–private partnerships are essential to the mission of HHS and ASPR because, speaking in military terms, HHS and ASPR do not have battalions or divisions of individuals at their command, he said. Rather, much of the nation's ability to protect its health in the face of pandemics, bioterrorism, and other large-scale events relies on public–private partnerships, with the federal government serving the role of convener and the source of some funds that can promote certain kinds of behaviors and activities, Kadlec explained.

As Kevin Yeskey stated in the workshop's opening presentation, the coalitions that have been created have demonstrated their effectiveness and shown they can make a tremendous difference in both preparedness and response. Similarly, said Kadlec, the small amount of critical infrastructure funding distributed through HPP made a significant difference during Hurricane Harvey to the patients and communities that hospitals serve. He added that the Ebola event drove home the lesson that there is no such thing as just-in-time preparedness.

Looking to the future, ASPR is focused on building the coalitions into a regional system, and Kadlec said he looks forward to receiving feedback from the community that will improve this program. His hope is that the community will see this program not as a top-down directive but as a means of nurturing and endorsing something implemented by communities at the local and regional levels. At the same time, he said this approach will only work if every organization and community is committed to it and believes it will be a benefit to participate in the coalitions.

From his perspective, he believes that preparedness has broad support in Congress from members of both parties who are committed to finding better ways of protecting the American public from a range of threats that are happening almost daily. "I would like to think that when I am done here, and more importantly when we are all collectively done here creating an enduring capability for our country, that we will look back and take a great deal of pride in knowing we are all committed to something that was bigger than ourselves and more enduring than we can imagine," said Kadlec. After putting in a plug for NDMS, he said that he and his colleagues are open to new ideas. He concluded his remarks by asking the participants to contact their representatives and senators to voice their support for the Pandemic and All-Hazards Preparedness Reauthorization Act.

DISCUSSION

Lewis Kaplan asked Kadlec how he envisions codifying the interactions between the federal government and the private sector so that information flow is bidirectional as well as dispersed. Kadlec replied that he is in the process of creating an organizational element in ASPR that will engage in outreach on a continual, rather than episodic, basis. He noted that BARDA already has this type of interaction codified as a regular part of its operations.

John Dreyzehner noted that he recently testified before the Senate Health Committee in support of the reauthorization. He made the point that in his experience, logistics and material are important, but that people and the relationships they form are really the safety net. Given the importance of relationships, he wondered how Kadlec plans to preserve existing relationships when the Strategic National Stockpile (SNS) completes its move from being part of CDC to being part of Kadlec's purview. Kadlec said that little will change other than the fact that the person in charge of the SNS will report to him instead of the CDC director. The purpose of the shift, he said, is to build a more coherent effort that should not affect the daily execution of the SNS, but should allow ASPR and CDC to work more closely together. He noted that starting April 1, 2018, he will have a representative from CDC in his office, and someone from ASPR will serve in the office of the CDC director. The plan is to expand ASPR's footprint with CDC at the regional and state levels, similar to the way that FEMA works with every state. "As we consolidate our efforts between ASPR and CDC, we hope to make [those interactions] more seamless," said Kadlec.

Thomas Kirsch asked Kadlec to comment on his vision for improved regional health system preparedness and response that will incorporate other federal assets, such as the VA and DoD facilities. Kadlec replied that during Hurricane Maria, national disaster medical assistance teams were stretched thin by the scope of the situation in Puerto Rico. Fortunately, there was a little-used section of the original Pandemic and All-Hazards Preparedness Act that allowed the VA to help, and the result was that the assistance teams saw 36,000 people in Puerto Rico and the VA saw 21,000 people, including VA beneficiary family members and the general public.

Going forward, ASPR has been working with the VA to expand the opportunity to use their physical facilities, personnel, logistics, electronic systems, and training to maximize its participation in large-scale disasters. Working with DoD is a little more challenging, he said, but he has several of his key deputies working with the special operations community at Fort Bragg to create opportunities to train and benefit from their experiences in Afghanistan and Iraq. The thing to remember, though, is that DoD's mission is overseas, not here, but given that, ASPR is trying to help DoD with preparing for the proposed Mission Zero, an amendment to the Public Health Service Act to facilitate assignment of military trauma care providers to civilian trauma centers in order to maintain military trauma readiness and to support such centers in times of need. He noted that during the December 2017 Amtrak derailment in Washington State, many of the injured were treated at Madigan Army Medical Center because it was embedded in the regional trauma coalition. On that note, the workshop was adjourned.

References

Auf der Heide, E., and J. Scanlon. 2007. The role of the health sector in planning and response. In *Emergency management: Principles and practice for local government*, 2nd ed., edited by W. J. Waugh, Jr., and K. Tierney. Washington, DC: ICMA Press.

Bachman, S. L., N. E. Demeter, G. G. Lee, R. V. Burke, T. W. Valente, and J. S. Upperman. 2014. The impact of trauma systems on disaster preparedness: A systematic review. *Clinical Pediatric Emergency Medicine* 15(4):296–308.

Berinato, S. 2003. All systems down. *Computerworld*, February 25, 2003.

Brown, J. B., M. R. Rosengart, T. R. Billiar, A. B. Peitzman, and J. L. Sperry. 2017. Distance matters: Effect of geographic trauma system resource organization on fatal motor vehicle collisions. *Journal of Trauma and Acute Care Surgery* 83(1):111–118.

Committee on Trauma and Trauma System Evaluation and Planning Committee. 2008. *Regional trauma systems: Optimal elements, integration, and assessment, American College of Surgeons Committee on Trauma: Systems consultation guide*. Chicago, IL: American College of Surgeons.

Cudnik, M. T., C. D. Newgard, M. R. Sayre, and S. M. Steinberg. 2009. Level I versus level II trauma centers: An outcomes-based assessment. *Journal of Trauma* 66(5):1321–1326.

DeSalvo, K. B. 2005. Letter from New Orleans. *Annals of Internal Medicine* 143(12):905–906.

DeSalvo, K. B. 2006. New Orleans healthcare after Katrina—1 year out. *Johns Hopkins Advanced Studies in Medicine* 6(7):305–306.

DeSalvo, K. B. 2016. New Orleans rises anew: Community health after Katrina. *Annals of Internal Medicine* 164(1):57–58.

DeSalvo, K. B. 2018. The health consequences of natural disasters in the United States: Progress, perils, and opportunity. *Annals of Internal Medicine* 168(6):440–441.

DeSalvo, K. B., and S. Kertesz. 2007. Creating a more resilient safety net for persons with chronic disease: Beyond the "medical home." *Journal of General Internal Medicine* 22(9):1377–1379.

Dobalian, A., R. Callis, and V. J. Davey. 2013. Evolution of the Veterans Health Administration's role in emergency management since September 11, 2001. *Disaster Medicine and Public Health Preparedness* 5(S2):S182–S184.

Guenther, R., and J. Balbus. 2014. *Primary protection: Enhancing health care resilience for a changing climate.* Washington, DC: Department of Health and Human Services.

Halamka, J. 2008. The caregroup network outage. In *Life as a Healthcare CIO.* Boston, MA: BlogSpot.

HRSA (Health Resources and Services Administration). 2006. *Model trauma system planning and evaluation.* Washington, DC: Department of Health and Human Services.

Kadlec, R. 2018. ASPR's new vision for a regional disaster health response system will help prepare nation for 21st century health security threats. In *ASPR Blog.* Washington, DC: Office of the Assistant Secretary for Preparedness and Response.

MacKenzie, E. J., F. P. Rivara, G. J. Jurkovich, A. B. Nathens, K. P. Frey, B. L. Egleston, D. S. Salkever, and D. O. Scharfstein. 2006. A national evaluation of the effect of trauma-center care on mortality. *New England Journal of Medicine* 354(4):366–378.

Maxson, T., C. D. Mabry, M. J. Sutherland, R. D. Robertson, J. O. Booker, T. Collins, H. J. Spencer, C. F. Rinker, T. L. Sanddal, and N. D. Sanddal. 2017. Does the institution of a statewide trauma system reduce preventable mortality and yield a positive return on investment for taxpayers? *Journal of the American College of Surgeons* 224(4):489–499.

NAS and NRC (National Academy of Sciences and National Research Council). 1966. *Accidental death and disability: The neglected disease of modern society.* Washington, DC: National Academy Press.

NASEM (National Academies of Sciences, Engineering, and Medicine). 2016. *A national trauma care system: Integrating military and civilian trauma systems to achieve zero preventable deaths after injury.* Washington, DC: The National Academies Press.

Appendix A

Workshop Agenda

March 20–21, 2018
National Academy of Sciences Building – Fred Kavli Auditorium
2101 Constitution Avenue, NW, Washington, DC 20418

MEETING OBJECTIVES

- Explore the degree to which the public and private health care systems self-identify as key components of the U.S. disasters and national security infrastructure;
- Discuss interest among health care institutions in developing collaborations across public and private sectors with the aim of strengthening capacity to respond to disasters and public health emergencies;
- Consider possible key levers that would motivate private-sector investment in system capacity building for disaster and public health emergency response, including those levers that already exist, but are not currently used as incentives to expand this capacity (quality measurement, grant programs, alternative payment models, tax benefits, etc.);
- Explore possible strategies to overcome key challenges to applying existing incentives to improve the quality, effectiveness, and efficiency of the management of critically ill and injured patients on a day-to-day basis and during emergency response scenarios;
- Review possible key sources of information and data elements that could be used to improve situational awareness of public- and private-sector health care facility capacity and capabilities to respond to disasters and public health emergencies; and
- Understand the degree to which Department of Defense or Department of Veterans Affairs hospitals could be used as a part of the U.S. response to disasters and public health emergencies requiring a health care response.

Day 1 – March 20, 2018

Session I	**Introduction and Overview of the Workshop**
8:30 AM	**Chairs' Welcome** HELEN BURSTIN Executive Vice President and Chief Executive Officer, Council of Medical Specialty Societies, and Co-Chair, Workshop Planning Committee ARTHUR L. KELLERMANN Dean and Professor, Edward F. Hebert School of Medicine, Uniformed Services University of the Health Sciences (USUHS), and Co-Chair, Workshop Planning Committee
8:45 AM	**Sponsor's Charge** KEVIN YESKEY Senior Advisor Office of the Assistant Secretary for Preparedness and Response (ASPR) Department of Health and Human Services (HHS)
Session II	**Perspectives on the National Capacity to Respond to Threats to Health, Safety, and Security**
9:15 AM	**Panel I: Private Health System Perspectives on the National Capability to Respond to Threats to Health, Safety, and Security** Moderator: • **Nicolette Louissaint**, Executive Director, Healthcare Ready Panelists: • **Brent James**, Senior Fellow, Institute for Healthcare Improvement, and Former Chief Quality Officer, Intermountain Healthcare • **John Perlin**, President, Clinical Services, and Chief Medical Officer, Hospital Corporation of America (HCA)

APPENDIX A

	• **Ronald Stewart**, Chair, Department of Surgery, University of Texas School of Medicine at San Antonio, and Chair, Executive Committee, Southwest Texas Regional Advisory Council (STRAC) • **David Witt**, Northern California Regional Chair, Infection Control, and Chair, Healthcare Continuity Clinical Workgroup, Kaiser Permanente
10:45 AM	Break
11:00 AM	**Panel II: Federal Perspectives on the National Capability to Respond to Threats to Health, Safety, and Security** Moderator: • **Thomas Kirsch**, Director, National Center for Disaster Medicine and Public Health, and Professor of Military and Emergency Medicine, F. Edward Hebert School of Medicine, USUHS Panelists: • **Kevin Hanretta**, Principal Deputy Assistant Secretary, Operations, Security and Preparedness, Department of Veterans Affairs (VA) • **Melissa Harvey**, Director, Division of National Healthcare Preparedness Programs, ASPR, HHS • **Anthony Macintyre**, Senior Medical Advisor and Medical Liaison Officer, Federal Emergency Management Agency (FEMA), Department of Homeland Security (DHS) • **Jody Wireman**, Director, Force Health Protection Division, North American Aerospace Defense Command Headquarters, U.S. Northern Command
12:15 PM	Lunch
Session III	**Learning from Past Experience**
1:15 PM	**Panel III: Case Studies in Cross-Sector Collaboration from Past Disruptions** Moderator: • **Ricardo Martinez**, Chief Medical Officer, Adeptus Health

Presenters:
- **Scott Cormier**, Vice President, Emergency Management, Environment of Care, and Safety, Medxcel Facilities Management
- **Karen DeSalvo**, Professor, Internal Medicine and Population Health, University of Texas; Former National Coordinator for Health Information Technology, HHS; and Former Commissioner, New Orleans Health Department
- **Erin Erb**, Division Vice President, Quality and Patient Safety, Gulf Coast Division, HCA
- **Todd Sklamberg**, Chief Executive Officer, Sunrise Hospital and Medical Center
- **Richard Zuschlag**, Chairman/Chief Executive Officer, Acadian Ambulance Service

3:15 PM **Break**

3:30 PM **Small Group Activity: ASPR's New Vision for a Regional Disaster Health Response System**

5:00 PM **Day One Wrap-Up**
HELEN BURSTIN
Executive Vice President and Chief Executive Officer, Council of Medical Specialty Societies, and Co-Chair, Workshop Planning Committee

ARTHUR L. KELLERMANN
Dean and Professor, Edward F. Hebert School of Medicine, USUHS, and Co-Chair, Workshop Planning Committee

5:30 PM **Adjourn Day One**

Day 2 – March 21, 2018

8:30 AM **Welcome and Recap of Day One**
HELEN BURSTIN
Executive Vice President and Chief Executive Officer, Council of Medical Specialty Societies, and Co-Chair, Workshop Planning Committee

ARTHUR L. KELLERMANN
Dean and Professor, Edward F. Hebert School of Medicine, USUHS, and Co-Chair, Workshop Planning Committee

Session IV **Looking Toward the Future**

8:45 AM **Panel IV: Cultivating Best Practices**

Moderator:
- **Skip Skivington,** Vice President, Healthcare Continuity Management and Support Services, Kaiser Permanente

Presenters:
- **Eileen Bulger,** Professor of Surgery, University of Washington; Chief of Trauma, Harborview Medical Center; and Chair, Committee on Trauma, American College of Surgeons
- **John Halamka,** Chief Information Officer, Beth Israel Deaconess Medical Center *(Remote Participant)*
- **James Jeng,** Professor of Surgery, Mount Sinai Healthcare System, and Chairman, Disaster Subcommittee, Committee on Organization and Delivery of Burn Care, American Burn Association
- **Gina Piazza,** Chief of Emergency Medicine, Charlie Norwood Medical Center, VA; Associate Professor, Emergency Medicine, Medical College of Georgia of Augusta University; and Co-Chair, High Threat Emergency Casualty Care Task Force, American College of Emergency Physicians (ACEP)

9:30 AM	**Panel V: Leading Change Across the Field**
	Moderator: • **Skip Skivington,** Vice President, Healthcare Continuity Management and Support Services, Kaiser Permanente
	Panelists: • **John Auerbach,** President/Chief Executive Officer, Trust for America's Health (TFAH), and Former Commissioner of Public Health, Commonwealth of Massachusetts • **Mitch Katz,** President/Chief Executive Officer, New York City Health and Hospitals, and Former Director, Los Angeles County Health Agency • **Ana McKee,** Executive Vice President and Chief Medical Officer, Joint Commission • **Craig Vanderwagen,** Partner, East West Protection, LLC, and Former ASPR, HHS
10:45 AM	**Break**
11:00 AM	**Panel VI: Leading Change at the Ground Level**
	Moderator: • **Jon Krohmer,** Director, Office of Emergency Medical Services, National Highway Traffic Safety Administration, Department of Transportation
	Panelists: • **Harold Engle,** Chief Nursing Officer and Healthcare Operations Executive, First Texas Cypress Fairbanks Medical Center Hospital • **Lewis Kaplan,** Associate Professor of Surgery, Perelman School of Medicine, University of Pennsylvania; Section Chief, Surgical Critical Care, Corporal Michael J. Crescenz Medical Center, VA; and Treasurer, Society of Critical Care Medicine • **Paul Kivela,** Managing Partner, Napa Valley Emergency Medical Group, and President, Board of Directors, ACEP

APPENDIX A 107

- **J. Brent Myers,** Chief Medical Officer, ESO Solutions, and Board President, National Association of EMS Physicians
- **Edward Racht,** Chief Medical Officer, American Medical Response, and Associate CMO, Evolution Health
- **Joseph Wright,** Professor and Chair, Department of Pediatrics and Child Health, Howard University College of Medicine

12:15 PM	**Lunch**
Session V	**Opportunities to Improve Private-Sector Investment in Capacity Building**
1:15 PM	**Panel VII: Regulatory Barriers and Facilitators**

Moderator:
- **Thomas Kirsch,** Director, National Center for Disaster Medicine and Public Health, and Professor of Military and Emergency Medicine, F. Edward Hebert School of Medicine, USUHS

Panelists:
- **Alex Camacho-Vásconez,** Regional Advisor, Emergency Preparedness and Disaster Risk Reduction, Pan American Health Organization (PAHO)
- **Duane Caneva,** Protective Medical Officer, Department of State
- **Sean Griffin,** Senior Manager, Policy and Coordination, North American Electric Reliability Corporation

2:15 PM	**Break**
2:30 PM	**Panel VIII: Financial Barriers and Facilitators**

Moderator:
- **Angela Brice-Smith,** Deputy Consortium Administrator for Quality Improvement and Survey & Certification Operations, Centers for Medicare & Medicaid Services

Panelists:
- **Andrew Baskin,** Vice President and National Medical Director for Quality, Aetna
- **David Frankford,** Professor of Law, Rutgers Law School; Professor, Rutgers Institute for Health, Health Care Policy and Aging Research; and Faculty Director, Rutgers Center for State Health Policy
- **Leslie Platt,** Senior Advisor, Health and Human Services, MITRE Corporation
- **Lucy Savitz,** Vice President, Research and Director, Center for Health Research (Oregon/Hawaii), Kaiser Permanente

Session VI	**Workshop Wrap-Up**	
3:30 PM	**Closing Keynote**	

ROBERT KADLEC
Assistant Secretary for Preparedness and Response, Department of Health and Human Services

4:00 PM **Closing Remarks**
HELEN BURSTIN
Executive Vice President and Chief Executive Officer, Council of Medical Specialty Societies, and Co-Chair, Workshop Planning Committee

ARTHUR L. KELLERMANN
Dean and Professor, Edward F. Hebert School of Medicine, USUHS, and Co-Chair, Workshop Planning Committee

4:30 PM **Adjourn Workshop**

Appendix B

Statement of Task

An ad hoc committee will organize and convene a 2-day public workshop in Washington, DC. Through this workshop, the committee will bring together public- and private-sector partners to discuss approaches to aligning health care system incentives with the American public's need for a health care system that is optimally prepared and scalable to manage acutely ill and injured patients during a disaster, public health emergency, or other mass casualty event. Specific topics that may be explored in this workshop include

- The degree to which the public and private health care systems self-identify as key components of the U.S. disasters and national security infrastructure;
- The interest among health care institutions in developing collaborations across public and private sectors with the aim of strengthening capacity to respond to disasters and public health emergencies;
- The possible key levers that would motivate private-sector investment in system capacity building for disaster and public health emergency response, including those levers that already exist, but are not currently used as incentives to expand this capacity (quality measurement, grant programs, alternative payment models, tax benefits, etc.);
- The possible strategies to overcome key challenges to applying existing incentives to improve the quality, effectiveness, and efficiency of the management of critically ill and injured patients on a day-to-day basis and during emergency response scenarios;

- The possible key sources of information and data elements that could be used to improve situational awareness of public- and private-sector health care facility capacity and capabilities to respond to disasters and public health emergencies; and
- The degree to which Department of Defense or Department of Veterans Affairs hospitals could be used as a part of the U.S. response to disasters and public health emergencies requiring a health care response.

The committee will develop the agenda for the workshop session, select and invite speakers and discussants, and moderate the discussions. Workshop proceedings will be prepared by a designated rapporteur in accordance with institutional guidelines, based on the presentations and discussions held during the workshop. The proceedings will be subject to appropriate review procedures before release.

Appendix C

Speaker Biographies

John Auerbach, M.B.A., is the President and CEO of the Trust for America's Health, a nonprofit, non-partisan organization dedicated to saving lives by protecting the health of every community and working to make disease prevention a national priority. He was formerly the Associate Director for Policy and the Acting Director of the Office for State, Tribal, Local and Territorial Support at the Centers for Disease Control and Prevention (CDC). As such, he managed CDC's Policy Office and oversaw key activities and technical assistance that supported the nation's health departments. Prior to his appointment at CDC, he was a Distinguished Professor of Practice in Health Sciences and the Director of the Institute on Urban Health Research and Practice at Northeastern University from 2012 to 2014. He was the Commissioner of Public Health for the Commonwealth of Massachusetts from 2007 to 2012. Prior to his appointment as Commissioner, Auerbach had been the Executive Director of the Boston Public Health Commission for 9 years.

Andrew Baskin, M.D., is Aetna's Vice President and National Medical Director for Quality. He works on initiatives to measure and improve quality of care, the provision of evidence-based care, quality measurement implementation and public reporting, health plan accreditation, and the establishment of performance based networks. Additionally, Dr. Baskin partners with others to help establish programs which create incentives for more effective and efficient care, influence and assure compliance with health care reform regulations, develop products to improve affordability and quality of care, and promote payment reform. Prior to this role, Dr. Baskin served in various medical director roles at Aetna, gaining experience and expertise in clinical and coverage policy development, clinical appeals, benefit and plan design, establishing coding and reimbursement policy, disease management program operations, and provider relations. Prior to joining Aetna in 1998, he practiced as a primary care Internal Medicine/Geriatrics physician in the Philadelphia suburbs. He is a member of the National Committee for Quality Assurance's (NCQA's) Committee on Performance Measurement, NCQA's Standards Committee, and a former two-term member of the National Quality Forum Consensus Standards Approval Committee.

Angela Brice-Smith, M.P.A., B.S.N., is Regional Administrator for the Atlanta & Dallas Regional Offices of the Centers for Medicare & Medicaid Services (CMS). She also serves as Deputy Consortium Administrator, Consortium for Quality Improvement and Survey & Certification Operations, at CMS. Ms. Brice-Smith has enjoyed more than 20 years of public service, and she is passionate about ensuring that beneficiaries, whether through the Medicare, Medicaid or Marketplace programs, get the finest and full value of the covered services needed for their conditions. Ms. Brice-Smith currently leads the Atlanta & Dallas Regional Offices' External Affair teams for outreach and information sharing with health care stakeholders, partners, providers, beneficiaries, and consumers. This activity has been critical in communicating key points on Medicare and Marketplace primarily, but also for the variety of special initiatives like efforts to address rural health, disparate health, engagement of more ethnic communities, or communications on Agency shifts in mission, for example shift to paying for quality health care rather than incentivizes toward payment of volume. Ms. Brice-Smith also provides management guidance and direction on survey activities involving health and safety concerns. Ms. Brice-Smith has worked in fee-for-service Medicare, Medicare managed care operations, and Medicare program integrity as lead and manager of areas related to Part A & B medical review, program oversight of Part B & Durable Medical Equipment, Prosthetics, Orthotics, and Supplies (DMEPOS) areas. She also served as the Deputy Director of Survey & Certification with notoriety in implementing user fees, initiating transplant, home health, hospice provisions, and inaugural systems. In the Medicaid arena, Ms. Brice-Smith was the Director of the Medicaid Integrity Group within the Center for Program Integrity. While there, she expanded the audit program and educational initiatives through collaborative efforts with states in several high fiscal risk, error-prone areas, and advanced efforts to reduce improper payments. She also led implementation of several provisions of the Patient Protection and Affordable Care Act involving payment suspension, use of Recovery Audit Contractors, as examples. Prior to joining CMS management, her non-management roles as a CMS health insurance specialist included being the subject-matter expert on marketing and regulatory requirements of managed care organizations. She worked as an analyst of Peer Review Organizations, when the agency first shifted to regional home health intermediaries for medical review of home health services. She worked in CMS's Office of the Actuary, manning the helpline for researcher inquiries, and provided analytical support to quasi review health care boards supporting CMS. Because Ms. Brice-Smith joined CMS (then the Health Care Financing Administration) as a Presidential Management Intern, she initiated rotations at the Business Group on Health, examining corporate health care issues; and in the Government Accountability Organization, as a researcher of the adequacy of the Public Health Service's budget in fighting acquired immunodeficiency syndrome. Prior to, and during her earlier years in CMS, Ms. Brice-Smith concurrently worked as a registered nurse. She worked in tertiary hospital Intensive Care Units, Emergency Departments, and even as a risk management/quality assurance coordinator in a Medicaid health maintenance organization. Once she joined CMS's management ranks, she left bedside nursing care. Ms. Brice-Smith received her Master's in Public Administration with focus in public finance, from Virginia Commonwealth University, and her Bachelor of Science in Nursing from the University of North Carolina at Chapel Hill.

Eileen Bulger, M.D., FACS, received her undergraduate degree from Johns Hopkins University and her medical degree from Cornell University. She then completed her surgical residency and fellowship in trauma and critical care at the University of Washington. She joined the faculty at the University of Washington in 2000 where she is now a Professor of Surgery and the Chief of Trauma at Harborview Medical Center the sole Level 1 trauma center for adults and pediatrics in Washington State. She has served as co–principal investigator for trauma with the Resuscitation Outcomes Consortium and PI of the Seattle Crash Injury Research and Engineering Network center. Her current research interests include resuscitation strategies for hemorrhagic shock, trauma system development and the prehospital triage of injured patients, the biomechanics of injury after motor vehicle crashes, modulation of the inflammatory response after injury, and necrotizing soft tissue infections. Her research has generated more than 200 peer-reviewed publications. She has been an active member of the National Disaster Medical System Trauma Critical Care Team-West since its inception in 2003. She has also served in several leadership roles in the Washington State trauma system including Chair of the State Emergency Medical Services and Trauma Steering committee, 2010–2015 and is a current member of the Washington State Disaster Medical Advisory Committee. She currently serves as the secretary/treasurer for the American Association for the Surgery of Trauma and the Chair of the American College of Surgeons Committee on Trauma.

Helen Burstin, M.D., M.P.H., FACP (*Co-Chair*), is the Executive Vice President and Chief Executive Officer of the Council of Medical Specialty Societies (CMSS). CMSS and its 43-member societies represent 790,000 U.S. physician members. CMSS member societies collaborate to enhance the quality of care delivered in the U.S. health care system and to improve the health of the public. Dr. Burstin formerly served as Chief Scientific Officer of the National Quality Forum (NQF), a not-for-profit membership organization that works to catalyze health care improvement through quality measurement and reporting. In her role, she was responsible for advancing the science of quality measurement and improvement. She is widely recognized for her work in patient-reported outcomes, risk adjustment, disparities, and patient safety. Prior to joining NQF, Dr. Burstin was the Director of the Center for Primary Care, Prevention, and Clinical Partnerships at the Agency for Healthcare Research and Quality (AHRQ). She led the development of the first National Healthcare Disparities Report and the use of practice-based research networks. Prior to joining AHRQ, Dr. Burstin was Director of Quality Measurement at Brigham and Women's Hospital and Assistant Professor at Harvard Medical School. She chaired the Quality Measures Workgroup of the Health IT Policy Committee. She was selected as a Baldrige Executive Fellow in 2016. Dr. Burstin is the author of more than 90 articles and book chapters on quality, safety, and disparities. Dr. Burstin is a graduate of the State University of New York at Upstate College of Medicine and the Harvard School of Public Health. She spent 1 year in Washington, DC, as National President of the American Medical Student Association. Dr. Burstin completed a residency in primary care internal medicine at Boston City Hospital. After residency, she completed fellowship training in General Internal Medicine and Health Services Research at Brigham and Women's Hospital and Harvard Medical School. She is a Professorial Lecturer in the Department of Health Policy at George Washington University School of Public Health and a Clinical Associate Professor of Medicine at George Washington University where she serves as a preceptor in internal medicine. She was awarded the Alpha Omega Alpha Medical Voluntary Attending Award from the George Washington School of Medicine.

Alex Camacho-Vásconez, M.D., currently serves as Regional Advisor on Emergency Preparedness and Disaster Risk Reduction in the Health Emergencies Department of the Pan American Health Organization based in Washington, DC. Dr. Camacho-Vásconez has more than 20 years of combined national and international experience in disaster risk reduction programs, including specific knowledge relating to the management and development of emergency medical services, pre-hospital care and management of complex emergencies and disasters at the national and international levels. He has been directly involved in the design of national and regional public policies for inclusive risk reduction and emergency medical services; the implementation of the strategy for Safe Hospitals in Ecuador's main public hospital; and leading the Ecuadorian Red Cross projects for humanitarian assistance during the eruption of Tungurahua Volcano, the heath sector response to AH1N1 pandemic influenza in Ecuador, and the Ecuadorian International Medical Team during the 2010 earthquakes in Haiti and Chile. He satisfactorily coordinated the implementation of Inclusive Disaster Risk Reduction project as National Secretary (Minister) of Disabilities of Ecuador. Dr. Camacho-Vásconez is a Medical Surgeon, holds a master degree in Health Management for Local Development and has completed additional studies related to health policies with scope on disasters. In the academic field, Dr. Camacho-Vásconez has taught at both the undergraduate and graduate levels in several Ecuadorian universities.

Scott Cormier serves as Vice President, Emergency Management, EC, and Safety, for Medxcel Facilities Management, a part of Ascension Healthcare. In this capacity, he oversees Emergency Management, Environment of Care, and Safety for the largest nonprofit health system in the United States. Mr. Cormier has led large system response to many disasters, and has published articles on hospital preparedness, emergency medical services, and influenza patient safety. Mr. Cormier is a past co-chair of the Healthcare and Public Health Sector Coordinating Council, a part of the federal critical infrastructure program, and currently serves as a past officer advisor. He also chairs the subcommittee for health care active shooter planning and response. He has more than 38 years of emergency management, public safety, and military experience.

Karen DeSalvo, M.D., M.P.H., M.Sc., is a physician whose career has been dedicated to improving the health of all people, with a particular focus on vulnerable populations, through patient care, education, policy and administrative roles, research, and public service. She is currently Professor of Medicine and Population Health at the University of Texas at Austin Dell Medical School. Dr. DeSalvo served as Acting Assistant Secretary for Health and also the National Coordinator for Health Information Technology at the Department of Health and Human Services during the Obama administration. She was previously New Orleans Health Commissioner and Vice Dean for Community Affairs and Health Policy at Tulane School of Medicine.

Harold Engle, M.B.A., R.N., CCRN-K, currently services as the Chief Nursing Officer (CNO) and Healthcare Operations Executive at the First Texas Cypress Fairbanks Hospital, a position he assumed in July 2016. As

CNO, Mr. Engle is among the leadership for a 50-bed inpatient hospital start-up in the Houston area, and he is an Emergency Preparedness leader for the hospital (he helped with the successful implementation of incident command during Hurricane Harvey). His supervision includes additional 29 satellite emergency rooms with a regional supervisory team. He performs the general functions of a CNO leading clinical program, and has developed the hospital's strategic growth plan in conjunction with the chief executive officer. Mr. Engle has developed new lines of service to increase revenues and managed expenses including but not limited to overtime control, nursing care hours per patient day, contracted services, and supplies. He has also utilized Studer® principles for customer experience and engagement and retention of staff. Mr. Engle supervises all nursing departments, laboratory, pharmacy, radiology, case management/social services, and surgery. Prior to his current role, Mr. Engle served in numerous related positions, including as a Neuro/cardiovascular Service Line Leader at Kingwood Medical Center; an Administrative Director of Emergency Services/Intensive Care Unit (ICU) at Memorial Hermann Memorial City; and a Director of Emergency Department/ICU/Cardiopulmonary/Therapy/Dialysis at St. Luke's Hospital at the Vintage. Since 2010, Mr. Engle has served as an associate faculty member at Ashford University. Mr. Engle received his Master's in Business Administration from Ashford University in 2009, his bachelor of sciences degree in nursing from Louisiana State University Medical Center School of Nursing in 1997, and his bachelor of sciences in microbiology from Louisiana State University A&M in 1993.

Erin Erb, M.H.S.A., R.H.I.A., C.P.H.Q., is the Division Vice President of Quality and Patient Safety at the Hospital Corporation of America (HCA) Gulf Coast Division. Ms. Erb is responsible for the overall clinical quality, patient outcomes, patient safety initiatives and emergency preparedness activities. Prior to joining the Gulf Coast Division Ms. Erb served in a various quality leadership roles at health systems in Nebraska and Kansas. Ms. Erb has overseen the implementation of high reliability organization principles, process improvement methodologies and has successfully overseen triennial accreditations and disease specific care certifications. Additionally, Ms. Erb has helped to drive significant improvement in patient facing metrics; decreased mortality, complications and hospital acquired infections. In August 2017, Ms. Erb and her colleagues at HCA provided incident command, response and recovery coordination to 19 hospitals in the Houston and Corpus Christi markets during Hurricane Harvey. Ms. Erb holds a bachelor's of science degree in Health Information Management and a master's degree in health care administration from the University of Kansas School of Medicine. She is certified through the American Health Information Association as a Registered Health Information Administrator, and a Certified Professional of Healthcare Quality by the National Association of Healthcare Quality.

David M. Frankford, J.D., is Professor of Law at Rutgers Law School; Professor at the Rutgers Institute for Health, Health Care Policy and Aging Research in New Brunswick; and Faculty Director at Camden of the Rutgers Center for State Health Policy. He has been a long-time editor of the *Journal of Health Politics, Policy and Law*, having served as book review and associate editor, the editor of "Behind the Jargon," a Special Section, and now a member of the Board of Editors. Professor Frankford's writings have focused on the interactions between health services research, health care politics and policy, and the institutions of professions and professionalism. His works include studies of state rate setting, hospital reimbursement, the regulation of fee splitting, the debates concerning privatization and national health insurance, the ideology of professionalism, the role of professionalism in medical education, the role of scientism and economism in health policy, issues of insurance coverage, and numerous other issues in health care financing. With Sara Rosenbaum, he is the author of the second edition of *Law and the American Health Care System*. He has been involved in many grants to the Rutgers Center for State Health Policy, offering analysis on such topics as state pharmacy assistance programs and hospital responses to mandatory medical error reporting. He also has participated in bioethics projects at The Hastings Center and the Center for Bioethics at the University of Pennsylvania. Currently his primary research interests concern the reconstitution of professionalism as the normative integration of professions and community, and the comparison of secular and religious bioethics regarding such issues as the new genetics.

Sean M. Griffin, M.A., is currently the Senior Manager for Policy and Coordination at the North American Electric Reliability Corporation and was previous Director for Incident Management Integration Policy at the White House National Security Council for President Obama and President Trump. At the White House, Mr. Griffin led the Executive Office of the President and interagency policy coordination for major disasters and incidents, including response and recovery to the 2016 Louisiana flood and Hurricane Matthew. Mr. Griffin led the effort to secure millions of dollars in Community Development Block Grant–Disaster Relief funds from Congress to aid the State of Louisiana following the historic flooding in 2016. Mr. Griffin is published on the whitehouse.gov blog where he provided valuable information to update American citizens on disaster response efforts, individual preparedness, and how to apply for Federal assistance as well as contribute to recovery efforts through donations and volunteering. Mr. Griffin chaired the Exercise and Evaluation Interagency Policy Committee leading to advancements in the National Exercise Program (NEP), leading the Senior Official Exercise Program, and directing the Presidential Transition Exercise Series to prepare the incoming Trump Administration Senior White House staff and Cabinet officials to respond to natural disasters, cyber incidents, infectious diseases, and terrorism. Mr. Griffin also led the revision to the NEP Base Plan, the policy that governs the NEP. Within the Office of Electricity Delivery and Energy Reliability at the Department of Energy (DOE), Mr. Griffin managed the DOE state energy assurance and exercise programs as well as solving critical policy and preparedness issues for energy sector resiliency and response as a member of industry-led and White House committees. Mr. Griffin directed and facilitated DOE's largest ever energy disaster exercise series, Clear Path, examining energy sector and cross-sector disaster response and recovery. Mr. Griffin was instrumental in coordinating power restoration efforts during the 2017 hurricane season for Texas, Florida, Puerto Rico, and the U.S. Virgin Islands. Mr. Griffin also led emergency management, training, and exercise programs at the Department of State, Defense Logistics Agency, and the National Institutes of Health. Mr. Griffin is also an active-duty veteran of the U.S. Navy and Naval Nuclear Power Program. Mr. Griffin volunteered his time to Chair the Federal Sector Emergency Managers Caucus for the International Association of Emergency Managers and is a leading member of the Private Sector Committee for the National Emergency Management Association. Mr. Griffin holds a Bachelor's in Nuclear Engineering Technology and completed a Master's in Emergency Management.

John D. Halamka, M.D., M.S., is Chief Information Officer of the Beth Israel Deaconess Medical Center, Chief Information Officer and Dean for Technology at Harvard Medical School, Chairman of the New England Health Electronic Data Interchange Network, chief executive officer of MA-SHARE (the Regional Health Information Organization), Chair of the U.S. Healthcare Information Technology Standards Panel (HITSP), and a practicing Emergency Physician. Dr. Halamka completed his undergraduate studies at Stanford University where he received a degree in Medical Microbiology and a degree in Public Policy with a focus on technology issues. While at Stanford he served as research assistant to Dr. Edward Teller, Dr. Milton Friedman, and presidential candidate John B. Anderson. He authored three books on technology related issues and formed a software development firm, Ibis Research Labs, Inc. Additionally, he served as a columnist for Infoworld, technical editor of *Computer Language Magazine* and technology consultant to several startup companies. In 1984, Dr. Halamka entered medical school at the University of California, San Francisco, and simultaneously pursued graduate work in Bioengineering at the University of California, Berkeley, focusing on technology issues in medicine. During medical school and graduate training, he continued his business activities and developed Ibis Research Labs into a 25 person software consultancy, specializing in medical and financial information interchange. Ibis was sold to senior management in 1992. Dr. Halamka served his residency at Harbor-University of California, Los Angeles (UCLA), Medical Center in the Department of Emergency Medicine. While at Harbor-UCLA he was an active member of the information systems team and developed a hospital-wide knowledge base for policies, procedures, and protocols. Furthermore, he was instrumental in creating an online medical record, a quality control system, and several systems for medical education. His research focus during residency was building automated triage tools for patients infected with human immunodeficiency virus. In 1996, Dr. Halamka joined the faculty of Harvard Medical School and continues to integrate his knowledge of medicine and technology focusing on the use of the Internet to exchange clinical patient data. His research includes security/confidentiality issues, scalability issues, and implementation of standards for exchange of administrative and clinical information. As a clinician as well as researcher, Dr. Halamka uses these tools to improve the care of the patients he treats in the Beth Israel Deaconess Emergency Department. He is also an active teacher, lecturing on both medical and technology topics to the students, residents, and faculty of Harvard and Massachusetts Institute of Technology. As Chief Information Officer at Beth Israel Deaconess, he is responsible for all clinical, financial, administrative and academic information technology serving 3,000 doctors, 12,000 employees, and 1 million patients. As Chief Information Officer and Dean for Technology at Harvard Medical School, he oversees all educational, research and administrative computing for 18,000 faculty and 3,000 students. As Chairman of NEHEN he oversees the administrative data exchange among the payers and providers in Massachusetts. As Chief Exchange Officer of MA-SHARE he oversees the Regional Healthcare Information Organization, which develops clinical data exchange efforts in Massachusetts. As Chair of HITSP he coordinates the process of electronic standards harmonization among stakeholders nationwide.

Kevin Hanretta, M.S., currently serves as Principal Deputy Assistant Secretary for Operations, Security, and Preparedness at the Department of Veterans Affairs (VA), a position he assumed in January 2017. He facilitates and integrates the VA's comprehensive emergency management "all-hazards" program as the VA's Acting Continuity Coordinator, directs personnel security programs, and law enforcement programs to ensure the Department can continue to perform mission essential functions under all circumstances across the threat spectrum. He has responsibility for the VA's 24/7 Integrated Operations Center, Continuity of Operations sites, and serves as the VA's Federal Senior Intelligence Coordinator. Before coming to the VA, Colonel Hanretta retired from the U.S. Army in 1999 with 30 years of service. He served as the Chief of Staff, Department of Defense 50th Anniversary of World War II Commemorations Committee. Colonel Hanretta served as a Ranger Advisor in Vietnam; with the 1st Ranger Battalion at Ft. Stewart, Georgia; commanded Headquarters Battery and B Battery, 319th Field Artillery in the 82nd Airborne Division, Ft. Bragg, North Carolina; activated and commanded the 37th Field Artillery Battalion in the 25th Infantry Division (Light), Hawaii; and served as Senior Aide to the Secretary of the Army. In 2004 he was detailed to the Department of Defense Iraq Transition Team in Baghdad and earned the Secretary for Defense Medal for Meritorious Civilian Service. In 2010, Colonel Hanretta received the Presidential Rank Award for meritorious exceptional career accomplishments and commitment to public service. Other positions he has held at the VA include Assistant Secretary, Office for Operations, Security, and Preparedness (2013–2017); Deputy Assistant Secretary for Emergency Management (2001–2006); and Deputy Chief of Staff (1999–2001). Colonel Hanretta received his Bachelor of Science degree in marketing from Siena College (1968)—where he also participated in U.S. Army Ranger Airborne, Pathfinder, and Air Assault Schools—and his Master of Science degree in logistics management from the Florida Institute of Techonology (1981). He also attended the Army War College (1990) and Leadership VA (2003). Most recently, Colonel Hanretta attended the Federal Executive Institute (2005, "Leadership in a Democratic Society"), the VA Senior Executive Strategic Leadership Course at the University of North Carolina (2010), and Federal Senior Intelligence Coordinator—Integrating the Intelligence Community (2015).

Melissa Harvey, M.S.P.H., R.N., is the Director of the Division of National Healthcare Preparedness Programs in the Office of the Assistant Secretary for Preparedness and Response (ASPR) at the Department of Health and Human Services. In this role, she is responsible for developing and advancing the implementation of policies and capabilities that aim to improve the nation's overall health care preparedness, including the Hospital Preparedness Program Cooperative Agreement. Recently, Ms. Harvey led ASPR's domestic health care system response to Ebola, including the development of a new regional and tiered strategy for the nation's health care facilities. She previously served as the Special Assistant to the ASPR, advising and supporting the Assistant Secretary on policy development, program implementation, and disaster response operations to ensure that the Office met its public health emergency preparedness and response mission. Ms. Harvey has also served as a Global Health Analyst, preparing assessments of foreign governments' capabilities to detect and respond to emerging infectious diseases, terrorism, and natural disasters. Prior to her work in the U.S. government, Ms. Harvey was the Program Manager of Emergency Management for the North Shore-Long Island Jewish Health System in New York, where she was responsible for all-hazards planning and response operations for the nation's second largest, nonprofit, secular health care system. She was also an emergency medical technician for the Health System's New York City 911 and inter-facility EMS divisions. Ms. Harvey attended Boston College, George Mason University, and Harvard University. She is a registered nurse in the Commonwealth of Virginia.

Brent James, M.D., M.S., is known internationally for his work in clinical quality improvement, patient safety, and the infrastructure that underlies successful improvement efforts, such as culture change, data systems, payment methods, and management roles. He is a member of the National Academy of Medicine, and participated in many of that organization's seminal works on quality and patient safety. He is a Fellow of the American College of Physician Executives. He holds faculty appointments at several universities: Clinical Professor, Stanford University School of Medicine (Medicine); Visiting Lecturer, Harvard School of Public Health (Health Policy and Management); Adjunct Professor, University of Utah David Eccles School of Business; and Adjunct Professor, University of Utah School of Medicine (Family Medicine; Biomedical Informatics). He is presently a Senior Fellow at the Institute for Healthcare Improvement (IHI), Boston, Massachusetts; a Senior Advisor at the Leavitt Group, Salt Lake City, Utah; and a Senior Advisor at Health Catalyst, Salt Lake City, Utah. He was formerly Chief Quality Officer and Executive Director at the Institute for Healthcare Delivery Research at Intermountain Healthcare, based in Salt Lake City, Utah. Through the Intermountain Advanced Training Program in Clinical Practice Improvement (ATP), he has personally trained more than 5,000 senior physician, nursing, and administrative executives, drawn from around the world, in clinical management methods, with proven improvement results (and leading to more than 50 "sister" training programs in more than 10 countries). He has been honored with a series of awards for quality in health care delivery, including: Distinguished Alumnus, University of Utah, 2015; Deming Cup–Columbia University School of Business, 2011; C. Jackson Grayson Medal, Distinguished Quality Pioneer–American Quality and Productivity Center, 2010; Joint Commission Ernest A. Codman Award, 2006; Health Research & Educational Trust TRUST Award, 2005; National Committee for Quality Assurance (NCQA) Quality Award, 2005; and American College of Medical Quality Founders' Award, 1999. For 8 of the first 9 years it existed, he was named among Modern Physician's "50 Most Influential Physician Executives in Healthcare." He was named among the "100 Most Powerful People in Healthcare" (Modern Healthcare) for 5 consecutive years, and was among Modern Healthcare's

"25 Top Clinical Informaticists." Before coming to Utah in 1986, he was Assistant Professor in the Department of Biostatistics at the Harvard School of Public Health, providing statistical support for the Eastern Cooperative Oncology Group (ECOG) and Cancer & Leukemia, Group B (CALG); and staffed the American College of Surgeons' Commission on Cancer. He holds Bachelor of Science degrees in Computer Science (Electrical Engineering) and Medical Biology; a Master of Statistics degree; and an M.D. (with residency training in general surgery and oncology). He serves on several nonprofit boards of trustees dedicated to clinical improvement and patient safety.

James Jeng, M.D., is a professor of surgery in the Mount Sinai Healthcare System (New York City) and serves as the Chairman of the Disaster Subcommittee, Committee on the Organization and Delivery of Burn Care, American Burn Association. Dr. Jeng has provided state of the art burn care for both run of the mill and extreme injuries in a three-state area of 7 million inhabitants (catchment area abutted burn centers at Johns Hopkins, Medical College of Virginia, and University of Pittsburg). For two decades, he has taught surgical trainees from Georgetown University, George Washington University, the U.S. Army and the U.S. Navy, and Howard University; teaching areas include trauma, acute care surgery, surgical critical care, burn care, and bench and translational research. Dr. Jeng became a recognized leader in the American burn community over two decades of working in diverse areas, including burn shock, end-points of burn shock resuscitation, laser applications in burn care, laser Doppler velocimetry and microvascular analysis, the National Burn Repository and data mining, data standards for burn care software, contingency planning for mass burn casualties, interface between the burn care community, the American Burn Association, key components of the federal government, nationwide situational awareness of burn care assets, uniform practice guidelines in burn care, and burn care under austere conditions. Internationally, Dr. Jeng is currently leading burn community efforts in burn disaster preparedness. In the role of International Society for Burn Injuries committee chairman, he helped launch a six-pronged methodology with deliverables aimed at 2016: (1) codify and diffuse knowledge on burn shock resuscitation using only oral fluids, (2) catalogue and diffuse knowledge of all known possible topical therapies for burn injuries, (3) systematically study and report on the phenomena/incidence of burn mass casualties around the globe so as to understand the extent of the problem, (4) continue efforts to bring further organization/connection between burn care providers and local governments, (5) catalogue, diffuse, and strengthen linkages between all NGOs with involvement in the worldwide burn care space, and (6) publish a multiauthor opinion piece in the journal to catalyze database development and data mining for burn injuries around the globe.

Robert Kadlec, M.D., M.S., is the Assistant Secretary for Preparedness and Response (ASPR) at the Department of Health and Human Services (HHS). The ASPR serves as the HHS Secretary's principal advisor on matters related to public health emergencies, including bioterrorism. The office leads the nation in preventing, responding to and recovering from the adverse health effects of manmade and naturally occurring disasters and public health emergencies. As such, the office coordinates interagency activities among HHS, other federal agencies, and state and local officials responsible for emergency preparedness and the protection of the civilian population from public health emergencies. Dr. Kadlec spent more than 20 years as a career officer and physician in the U.S. Air Force before retiring as a Colonel. Over the course of his career, he has held senior positions in the White House, the U.S. Senate, and the Department of Defense (DoD). Most recently, he served as the Deputy Staff Director to the Senate Select Committee on Intelligence. Dr. Kadlec previously served as staff director for Senator Richard Burr's subcommittee on bioterrorism and public health in the 109th Congress. In that capacity, he was instrumental in drafting the Pandemic and All-Hazard Preparedness Bill which was signed into law to improve the nation's public health and medical preparedness and response capabilities for emergencies, whether deliberate, accidental, or natural. Dr. Kadlec also served at the White House from 2002 to 2005 as director for biodefense on the Homeland Security Council, where he was responsible for conducting the biodefense end-to-end assessment, which culminated in drafting the National Biodefense Policy for the 21st Century. He served as Special Assistant to President George W. Bush for Biodefense Policy from 2007 to 2009. Earlier in his career, he served as the Special Advisor for Counter Proliferation Policy at the Office of the Secretary of Defense, where he assisted DoD efforts to counter chemical, biological, radiological, and nuclear (CBRN) threats in the wake of 9/11 and contributed to the Federal Bureau of Investigation's investigation of the anthrax letter attacks. He began his career as a flight surgeon for the 16th Special Operations Wing and subsequently served as a surgeon for the 24th Special Tactics Squadron and as Special Assistant to J-2 for Chemical and Biological Warfare at the Joint Special Operations Command. He was named U.S. Air Force Flight Surgeon of the Year in 1986. Dr. Kadlec holds a bachelor's degree from the U.S. Air Force Academy, a doctorate of medicine and a master's degree in tropical medicine and hygiene from the Uniformed Services University of the Health Sciences, as well as a master's degree in national security studies from Georgetown University.

Lewis Kaplan, M.D., FACS, FCCM, FCCP, is a general, trauma, and critical care surgeon at the Perelman School of Medicine at the University of Pennsylvania (where he is an Associate Professor of Surgery) who serves as the Section Chief of Surgical Critical Care at the Corporal Michael J. Crescenz VA Medical Center in Philadelphia, Pennsylvania. He received a B.A. in Biology with honors from Franklin and Marshall College in Lancaster, Pennsylvania, in 1980 and his medical degree from Rutgers Medical School in Piscataway, New Jersey, in 1984. Dr. Kaplan completed his surgical residency at the Medical College of Pennsylvania (MCP; 1988–1995) with 2 years spent in basic research into cardiac bioenergetics and ischemic preconditioning (1991–1993). Dr. Kaplan completed a Fellowship in Surgical Critical Care at the University of Pittsburgh Medical Center (1996–1997) and then joined the faculty at MCP and Hahnemann University Hospital in Philadelphia, where he directed the surgical intensive care unit (ICU) and the Surgical Critical Care (SCC) fellowship. Seven years later he was recruited to Yale University to establish an Emergency General Surgery service for the Division of Trauma and Surgical Critical Care. Having done so he then resumed leadership in the Yale-New Haven Hospital ICU and the SCC and Acute Care Surgery fellowships. Eleven years after that, he was recruited back to Philadelphia into his current roles. Dr. Kaplan serves in several professional societies in leadership roles (Society of Critical Care Medicine, Surgical Infection Society), multiple editorial boards including the *Journal of Trauma and Acute Care Surgery*, *Surgical Infections*, and reviews for a host of others. Dr. Kaplan's durable interest in Tactical Emergency Medical Services is underscored by his serving for years as a surgeon embedded in a regional special weapons and tactics team. Dr. Kaplan's research interests span unmeasured ion impact in acid-base balance, acute kidney injury, surgical infection, and Tactical Emergency Medical Services.

Mitchell H. Katz, M.D., is currently President and Chief Executive Officer at New York City Health + Hospitals. Dr. Katz is a highly experienced public health care executive and physician with a track record of achieving measurable results throughout his career. Previously, Dr. Katz was the Director of the Los Angeles County Health Agency, an agency that combines the Departments of Health Services, Public Health, and Mental Health into a single entity so as to provide more integrated care and programming within Los Angeles. The Agency has a $7 billion budget, 28,000 employees, and a large number of community partners. Dr. Katz served as the Director of the Los Angeles County Department of Health Services (DHS), the second largest public safety net system in the United States. During this time, he created the ambulatory care network and empaneled more than 350,000 patients to a primary care home. He eliminated the deficit of DHS through increased revenues and decreased administrative expenses, and used the new Patient Protection and Affordable Care Act funding to pay for a modern electronic health system, Orchid, which has now been implemented in 90 percent of DHS clinical sites. He moved more than 1,000 medically complex patients from hospitals and emergency departments into independent housing, thereby eliminating unnecessary expensive hospital care and giving the patients the dignity of their own home. Dr. Katz continued to see patients every week as an outpatient physician at Edward R. Royal Comprehensive Health Center and on the inpatient medicine service at Los Angeles County+University of Southern California Medical Center (LAC+USC), Harbor-University of California, Los Angeles (UCLA), and Olive View-UCLA Medical Centers. Before he came to Los Angeles, Dr. Katz was the Director and Health Officer of the San Francisco Department of Health for 13 years. He is well known for funding needle exchange, creating Healthy San Francisco, outlawing the sale of tobacco at pharmacies, and winning ballot measures for rebuilding Laguna Honda Hospital and San Francisco General Hospital. He is a graduate of Yale College and Harvard Medical School. He completed an internal medicine residency at University of California, San Francisco, Medical School and was an Robert Wood Johnson Clinical Scholar. He is the Deputy Editor of the *Journal of the American Medical Association (JAMA) Internal Medicine*, an elected member of the National Academy of Medicine, and the recipient of the Los Angeles County Medical Association 2015 Healthcare Champion of the Year award.

Arthur L. Kellermann, M.D., M.P.H. (*Co-Chair*), became Dean of the F. Edward Hébert School of Medicine at the Uniformed Services University of the Health Sciences (USUHS), on September 7, 2013. The unique program has ranked among the top in the nation, and is the country's only federal medical school. Dr. Kellermann's distinguished career is anchored in academic medicine and public health. Prior to joining USUHS, he held the Paul O'Neill-Alcoa Chair in Policy Analysis at RAND, a nonprofit research organization. He was a professor of emergency medicine and public health and associate dean for health policy at the Emory School of Medicine in Atlanta. He founded Emory's Department of Emergency Medicine and served as its first chair from 1999 to 2007. He also founded the Emory Center for Injury Control, a World Health Organization Collaborating Center. A two-term member of the board of directors of the American College of Emergency Physicians, Dr. Kellermann was subsequently given the College's highest award for leadership. He also holds "excellence in science" awards from the Society for Academic Emergency Medicine and the Injury Control and Emergency Health Services Section of the American Public Health Association. Elected to the National Academy of Medicine (NAM) in 1999, he co-chaired the Committee on the Consequences of Uninsurance. He currently serves on the NAM's Governing Council. A clinician and researcher, Dr. Kellermann practiced and taught emergency medicine for more than 25 years in public teaching hospitals in Seattle, Washington; Memphis, Tennessee; and Atlanta, Georgia. His research addresses a wide range of issues, including health care spending and information technology, injury prevention, treatment of traumatic brain injury, emergency care and disaster preparedness.

Thomas Kirsch, M.D., M.P.H., is the Director of the National Center for Disaster Medicine and Public Health (NCDMPH) and a Professor of Military and Emergency Medicine at the Uniformed Services University of the Health Sciences. He is a board-certified emergency physician and expert in disaster management and science, austere medicine and health care management. He comes to NCDMPH from Johns Hopkins University as a Professor of Emergency Medicine, International Health and Civil Engineering. He has responded to many national and global disasters including hurricanes Katrina (2005) and Sandy (2012), the New York City response to the 9/11 terrorist attacks (2001), global disasters such as the earthquakes in Haiti (2010), Chile (2010), and New Zealand (2011), and the 2010 floods in Pakistan and Typhoon Haiyan in the Philippines (2013). He has consulted on disaster and humanitarian related issues for organizations such as the Centers for Disease Control and Prevention, Federal Emergency Management Agency, the Department of Defense, Office of Foreign Disaster Assistance, the American and Canadian Red Cross, the World Health Organization, the United Nations Children's Fund, Pan American Health Organization, and the Earthquake Engineering Research Institute. Dr. Kirsch is a globally recognized teacher who has lectured extensively nationally and internationally on disaster and emergency medicine issues. While at Johns Hopkins he founded and was the Director of the Johns Hopkins School of Medicine Austere Medicine and the Disaster Medicine Fellowship. He has also taught masters and doctorate-level courses in the Hopkins School of Public Health and School of Medicine. Dr. Kirsch has authored more than 100 scientific articles, abstracts, and textbook chapters, and co-authored the austere medical textbook, Emergent Field Medicine (VanRooyen-Kirsch). In 2013 he received the inaugural Disaster Science Award from the American College of Emergency Physicians and in 2014 the Clara Barton Award for Leadership from the American Red Cross. He was also recognized as a Hero in Healthcare Fighting Ebola by President Obama in a White House ceremony in 2014. He received a B.A. in Fine Arts from Creighton University, his M.D. from the University of Nebraska, and an M.P.H. from the Johns Hopkins School of Public Health, and then completed an Emergency Medicine Residency at the George Washington-Georgetown Combined Program.

Paul Kivela, M.D., M.B.A., FACEP, is a residency-trained and board certified emergency physician who works clinically as managing partner of the Napa Valley Emergency Medical Group, a single hospital democratic group in Napa, California. He is currently President of the American College of Emergency Physicians. In the last several years, Napa has taken part in the response to multiple disasters, including an earthquake, multiple large fires, and multiple shootings. Dr. Kivela has several leadership positions including Medical Director of Medic Ambulance, Vice President of FailSafe Healthcare where he develops risk management and patient safety solutions, and advisor to several small and midsize groups. He works to develop innovative solutions and has negotiated successfully with both insurance companies and employer based plans. Dr. Kivela is a graduate of the University of Illinois where he also earned his medical degree. He completed his emergency medicine residency at the University of California, Los Angeles. He received his M.B.A. at the University of Tennessee.

Jon Krohmer, M.D., is the Director of the Office of Emergency Medical Services (OEMS) at the National Highway Traffic Safety Administration (NHTSA), a position in which he has served since 2016. Dr. Krohmer has a wealth of emergency medical services experience and expertise and leads NHTSA's collaborative efforts to improve emergency care across the nation. Board certified as an emergency physician, Dr. Krohmer has been actively involved in EMS for more than 30 years, first in his home state of Michigan, and then at the national level, as an active member of the American College of Emergency Physicians (ACEP) and president of the National Association of EMS Physicians (NAEMSP). Dr. Krohmer's federal service began as the Deputy Assistant Secretary for Health Affairs and Deputy Chief Medical Officer for the Department of Homeland Security (DHS). He went on to become the Director of the Health Services Corps for Immigration and Customs Enforcement at DHS before joining NHTSA, and recently he provided medical expertise to the U.S. Coast Guard as well.

Nicolette A. Louissaint, Ph.D., serves Healthcare Ready as the Director of Programming. In this role, she manages all of the programs and policies related to emergency response and health care operations. Most recently, Dr. Louissaint served as a Foreign Affairs Officer at the Department of State in the Bureau of Economic and Business Affairs as the lead officer for health intellectual property and trade issues. During the height of the Ebola Epidemic of 2014, Dr. Louissaint served as the Senior Advisor to Ambassadors Nancy J. Powell and Steven A. Browning, the Department of State's Special Coordinators for Ebola. She holds degrees in Chemical Engineering and Biological Sciences from Carnegie Mellon University, as well as a Ph.D. in

Pharmacology and Molecular Sciences from the Johns Hopkins University School of Medicine.

Anthony Macintyre, M.D., serves as the Department of Homeland Security (DHS) Office of Health Affairs (OHA) Senior Medical Advisor and Medical Liaison Officer to the Federal Emergency Management Agency (FEMA). He is a Board Certified Emergency Physician still practicing clinically. He is a Clinical Professor with the Department of Emergency Medicine at George Washington University. His academic and field response careers have focused on medical emergencies and disasters at multiple levels. Since 1995, Dr. Macintyre has served as the Medical Director for Fairfax County's Urban Search and Rescue Program, with deployments to numerous domestic incidents including the bombing of the Murrah Building in Oklahoma City (1995), the Pentagon terrorist attack (2001), and numerous hurricanes including Hurricane Katrina–Mississippi area of operations (2005). His international deployments have included responding to the bombing of the U.S. Embassy in Nairobi (1998) and devastating earthquakes in Turkey (1999), Taiwan (1999), Iran (2003), Haiti (2010), and Nepal (2015). These deployments were part of the official U.S. Government rescue efforts through the U.S. Agency for International Development/Office of U.S. Foreign Disaster Assistance (USAID/OFDA). Dr. Macintyre has assisted other agencies such as the Veterans Health Administration (VHA), Department of State, USAID/OFDA, and the Department of Health and Human Services (National Disaster Medical System) with medical emergency planning and response efforts. In 2002, Dr. Macintyre served as an assistant investigator to Dr. Joseph Barbera in the Alfred P. Sloan Foundation–funded project to develop the Medical and Health Incident Management system (MaHIMs). This product provides a comprehensive, functionally-based model for the response to and management of complex, large-scale medical emergencies. He co-authored *Medical Surge Capacity and Capability: A Management System for Integrating Medical and Health Resources during Large-Scale Emergencies*. This book is currently used as a template by the Department of Health and Human Services (HHS) Hospital Preparedness Program (HPP). Dr. Macintyre co-developed a mass decontamination capability for the old George Washington University Hospital and published key concepts in the *Journal of the American Medical Association*. As an emergency physician, he was instrumental in structuring the hospital response to the 2001 anthrax dissemination incident. Dr. Macintyre served for more than a decade on the District of Columbia Hospital Association (DCHA) Emergency Preparedness Committee and its successor, the Emergency Management Committee for the DC Emergency Healthcare Coalition (DC EHC)—both focused on enhancing the health and medical response to disasters in Washington, DC.

Ricardo Martinez, M.D., FACEP, is a nationally recognized board-certified emergency physician and has practiced emergency medicine clinically for more than 30 years, and held senior roles in business, academia, and the federal government. He currently serves as the Chief Medical Officer for Adeptus Health. Before joining Adeptus Health, Dr. Martinez was Chief Medical Officer of North Highland Worldwide Consulting, where a major focus of his work was collaborating with physician leadership to enhance their effectiveness in providing high-value care, building data-driven patient-centered teams, and driving cultural change. Dr. Martinez also served as the Executive Vice President of Medical Affairs for the Schumacher Group, a leading emergency medicine practice management company, and was previously appointed Federal Administrator of the National Highway Traffic Safety Administration (NHTSA) by President Bill Clinton. He currently serves as faculty at Emory University School of Medicine and previously held roles at Stanford University School of Medicine and as Executive Director of the Medical Leadership Academy. Dr. Martinez has been a senior medical advisor to the National Football League since 1988, facilitating medical care, emergency planning, preparedness, and public health for the Super Bowl. He was elected to the National Academy of Medicine in 2004 and served on the Board of Directors of the Public Health Foundation. Dr. Martinez pursued undergraduate studies from Louisiana State University, an M.D. from Louisiana State University School of Medicine, and his residency at Louisiana State University-Charity Hospital at New Orleans, where he was Chief Resident.

Ana Pujols McKee, M.D., is the Executive Vice President and Chief Medical Officer of the Joint Commission. In this role, Dr. McKee represents the Joint Commission enterprise as she focuses on and develops policies and strategies for promoting patient safety and quality improvement in health care. Her responsibilities include providing support to the Joint Commission's Patient Safety Advisory Group; overseeing work related to the development of the Sentinel Event Policy, National Patient Safety Goals and Sentinel Event Alerts; supervising the Sentinel Event Database; and overseeing the functions of the Standards Interpretation Group and the Office of Quality and Patient Safety. Dr. McKee is the former board Chair of the Pennsylvania Patient Safety Authority and former Vice Chair of the Philadelphia Public Health Management Corporation. Dr. McKee has also served as a board member for the American Cancer Society, the Pennsylvania Health Care Cost Containment Council, Health Partners Philadelphia, the Philadelphia AIDS Consortium, and Quality Insights of Pennsylvania. In addition, she served on the Food and Drug Administration's Advisory Committee and on several committees of the National Institutes of Health. Dr. McKee holds a bachelor's degree in psychology from the State University of New York at Binghamton and a medical degree from Hahnemann Medical College and Hospital in Philadelphia. She completed her residency at Presbyterian Medical Center in Philadelphia, and is board certified in internal medicine. Among her most recent acknowledgments, Dr. McKee was recognized among the Top 25 Minority Executives in Healthcare for 2014.

Jonathan Brent Myers, M.D., M.P.H., in an internationally recognized expert in the areas of emergency medical services (EMS), clinical informatics, and population health. He serves as Chief Medical Officer for ESO Solutions and Associate Medical Director for Wake EMS in Raleigh, North Carolina as well as President of the National Association of EMS Physicians (NAEMSP). He served as co-editor of the 2nd Edition of *Emergency Medical Services, Clinical Practice and Systems Overnight*, serves on the editorial board of numerous peer-reviewed journals, has many peer-review publications, and has presented more than 120 invited lectures, both nationally and internationally. He maintains triple board certification in the areas of Emergency Medicine, EMS, and Clinical Informatics with a focus on consistent implementation of evidence-based techniques to improve the health of populations.

Jonathan B. Perlin, M.D., Ph.D., M.S.H.A., M.A.C.P., FACMI, is president, clinical services and chief medical officer of the Nashville, Tennessee-based Hospital Corporation of America (HCA). He provides leadership for clinical services and improving performance at HCA's 177 hospitals and more than 1,000 outpatient surgical, urgent care and other practice units. Current activities include advancing electronic health records for learning health care and continuous improvement; driving value through (big) data science and advanced analytics; and elevating measured clinical performance and patient safety to benchmark levels. His team recently completed the landmark REDUCE MRSA study (Randomized Evaluation of Decolonization vs. Universal Clearance to Eliminate methicillin-resistant *Staphylococcus aureus* [MRSA]) that demonstrated a 44 percent improvement on known best practices for reducing bloodstream infections. Before joining HCA in 2006, "the Honorable Jonathan B. Perlin" was Under Secretary for Health in the Department of Veterans Affairs. Nominated by President George W. Bush and confirmed by the Senate, as the senior-most physician in the Federal Government and Chief Executive Officer of the Veterans Health Administration (VHA), Dr. Perlin led the nation's largest integrated health system. At the VHA, Dr. Perlin directed care to more than 5.4 million patients annually by more than 200,000 health care professionals at 1,400 sites, including hospitals, clinics, nursing homes, counseling centers and other facilities, with an operating and capital budget of $37.4 billion. A champion for early implementation of electronic health records, Dr. Perlin led VHA quality performance to international recognition as reported in academic literature and lay press and as evaluated by RAND, the Institute of Medicine, and others. Dr. Perlin was the 2015 chairman of the American Hospital Association. He also serves as chair of the Secretary of Veterans Affairs Special Medical Advisory Group. From July to September 2014 Dr. Perlin took a sabbatical to serve as Senior Advisor to the Secretary of Veterans Affairs to help improve operations, accelerate access and rebuild trust with America's Veterans. Dr. Perlin has served previously on numerous Boards and Commissions including the Joint Commission and the National Patient Safety Foundation and currently serves on the Board of Meharry Medical College and the National Quality Forum. He was the inaugural chair of the Department of Health and Human Services Health IT Standards Committee. A member of the National Academy of Medicine and recognized perennially as one of the most influential physician executives and health leaders in the United States by Modern Healthcare, Dr. Perlin has received numerous awards including Distinguished Alumnus in Medicine and Health Administration from his alma mater, Chairman's Medal from the National Patient Safety Foundation, the Founders Medal from the Association of Military Surgeons of the United States, and is one of the few honorary members of the Special Forces Association and Green Berets. Broadly published in health care quality and transformation, Dr. Perlin is a Master of the American College of Physicians and Fellow of the American College of Medical Informatics. He has a Master's of Science in Health Administration and received his Ph.D. in pharmacology (molecular neurobiology) with his M.D. as part of the Physician Scientist Training Program at the Medical College of Virginia of Virginia Commonwealth University (VCU). Dr. Perlin has faculty appointments at Vanderbilt University as Clinical Professor of Medicine and Biomedical Informatics and at VCU as Adjunct Professor of Health Administration.

Gina Piazza, D.O., FACEP, is the Chief of Emergency Medicine at the Charlie Norwood Department of Veterans Affairs Medical Center and an Associate Professor of Emergency Medicine at the Medical College of Georgia of Augusta University in Augusta, Georgia. She completed her residency in Emergency Medicine at Lincoln Medical and Mental Health Center in New York City and an EMS/Disaster Medicine Fellowship through the State University of New York at Buffalo. She also completed a Health Care Policy Fellowship at the Emergency Care Coordination Center within the Department of Health and Human Services. Dr. Piazza has served as a tactical physician for local and federal law enforcement agencies, and has taught tactical medicine, hospital emergency management, and disaster medicine throughout the United States and abroad. She currently serves as the co-chair of the American College of Emergency Physicians' High Threat Emergency Casualty Care Task Force.

Leslie Platt, J.D., is a Senior Advisor on Health and Human Services at the MITRE Corporation. Mr. Platt is a nationally recognized life sciences attorney and executive with experience at senior levels in government and industry on national health policy, biomedical research, development, and commercialization; program funding, financing, and impact investment strategies; public–private partnerships; and many other areas. In the Federal government, Mr. Platt was a Charter Member of the federal Senior Executive Service and advised Cabinet officers and top government officials on a range of high priority legal, policy, and management issues. Mr. Platt served as Executive Assistant to the Director and Chief of Operations, Office of the Director, at the National Institutes of Health; Deputy General Counsel-Legal Counsel at the Department of Health and Human Services; Counsel and Staff Director of the White House Agent Orange Working Group; and earlier, as Associate General Counsel for Legislation at the Department of Housing and Urban Development. In the private sector, Mr. Platt has advised Boards of Directors and C-Suites of public and closely held corporations and nonprofit organizations, as Counsel at Pillsbury Winthrop Shaw Pittman LLP; Principal at Ernst & Young LLP; Chief Operating Office & General Counsel at The Institute for Genomic Research; and Senior Vice President & General Counsel at the ATCC, and in other positions. Mr. Platt has received many awards for distinguished public service and leadership, has authored numerous articles and taught at the graduate level, and has been a frequent speaker at industry conferences in the United States and around the world.

Edward M. Racht, M.D., is the Chief Medical Officer for American Medical Response and Associate Chief Medical Officer of Evolution Health. Dr. Racht has more than 20 years of experience in emergency medical services and health care systems. Previously, Dr. Racht served as the Chief Medical Officer and Vice President of Medical Affairs for Piedmont Newnan Hospital in metro Atlanta and as Medical Director for the Austin/Travis County Emergency Medical Services (EMS) System in Texas, which was nationally recognized for its collaborative approach to resolving challenging health care integration issues. Dr. Racht has served three successive terms on the Virginia State Governor's EMS Advisory Board and chaired the state of Texas Governor's EMS and Trauma Advisory Council for 10 years. He received his undergraduate and medical degrees from Emory University in Atlanta and completed his residency at the Medical College of Virginia.

Lucy A. Savitz, Ph.D., M.B.A., has more than 30 years of experience in health care delivery and health services research. Currently, she is Vice President, Health Research for Kaiser Permanente (KP) Northwest Region and Director for the KP Centers for Health Research in Oregon and

Hawaii. Dr. Savitz has led numerous implementation and evaluation studies over her 30-year career with a focus on quality, safety, and elimination of unwarranted variation (i.e., waste). Furthermore, she has been acknowledged as an Examiner for the 2001 and 2002 Malcolm Baldrige National Quality Program, administered by the National Institute of Standards and Technology in the Department of Commerce and the American Society for Quality. She is faculty for the Institute for Healthcare Improvement (IHI). Additionally, she holds numerous academic appointments—affiliate Professor, Health Systems Management & Policy at the Oregon Health & Sciences University-Portland State University School of Public Health; Adjunct Research Professor, Clinical Epidemiology at the University of Utah School of Medicine; and Adjunct Associate Professor, Health Policy & Management at the University of North Carolina at Chapel Hill Gillings School of Public Health. At Academy Health, she serves as Chair on the Methods & Data Council, a member of the Delivery System Science Fellowship Program Committee, and Chair of the Committee for Advocacy in Public Policy. For the Centers for Medicare & Medicaid Services, she is an invited member of the Executive Leadership Council; and at the Agency for Healthcare Research and Quality, serves on the National Advisory Council. Her current area of thought leadership internationally as well as within KP focuses on evolving the methods and metrics needed to accelerate implementation of safety interventions and realistic program evaluations that support the learning health system.

Skip Skivington, M.B.A., has worked at Kaiser Permanente for more than 16 years and is currently the Vice President, Healthcare Continuity & Support Services, based at the national headquarters in Oakland, California. Mr. Skivington is responsible for the executive oversight of Kaiser Permanente's Supplier Diversity, Nutritional Services, Corporate Meeting Services, Materials Management, Product Recall, Vendor Authorization, and Healthcare Continuity Management programs. Since 2000, Mr. Skivington has been responsible for the implementation of a formal health care continuity management program throughout Kaiser Permanente. In addition to directing this formal planning process, and immediately following the anthrax attacks in October 2001, Mr. Skivington formed and now directs Kaiser Permanente's threat assessment program consisting of an executive oversight council and functional working groups in the disciplines of clinical (physicians, nursing and laboratories), facilities, community linkages, people, legal, communications and education, supply chain, and public policy. Mr. Skivington is a member of the State of California Joint Advisory Committee for terrorism preparedness, and the American Health Insurance Plans' Disaster Readiness Committee. Mr. Skivington is also a member of the Conference Board's Business Continuity and Crisis Management

Council, the U.S. Healthcare Sector Critical Infrastructure Council, and is a frequent speaker on medical preparedness in the event of a terrorist attack. Following Hurricanes Katrina and Rita, Mr. Skivington led two Kaiser Permanente medical response teams consisting of physicians, nurses and mental health providers to the Gulf Region at the request of the U.S. Surgeon General and the State of California. Mr. Skivington was the project administrator for the U.S. Government's Hospital Incident Command System (HICS) Revision IV Project. HICS IV was updated on behalf of the government through a national working group representing hospitals throughout the country, along with input from national agencies including the American Hospital Association, the Joint Commission, the Health Resources and Services Administration, the Department of Health and Human Services, and the Federal Emergency Management Agency. Mr. Skivington holds both a B.A. and an M.B.A. in Business Administration.

Todd Sklamberg, M.B.A., is the Chief Executive Officer of Sunrise Hospital and Medical Center, Nevada's largest acute care facility at 690 beds, and Sunrise Children's Hospital located on the same campus. Prior to becoming CEO, Mr. Sklamberg served as the Chief Operating Officer of Sunrise Children's Hospital—Nevada's largest, most comprehensive children's hospital. Sunrise Hospital and Sunrise Children's Hospital are members of the Hospital Corporation of America's (HCA's) Sunrise Health System in Las Vegas. Prior to joining HCA, Mr. Sklamberg served as vice president at St. Louis Children's Hospital. Mr. Sklamberg currently serves as a member of the Executive Committee for the Nevada Hospital Association. He is also active with the Las Vegas Metro Chamber of Commerce serving as co-chair of the Healthcare Policy Task Force and is a member of the Government Affairs Committee. His community involvement also includes committee work with the Regional Transportation Commission including the Transport Resource Advisory Committee, Maryland Parkway Business Coalition and Maryland Parkway Corridor Committee. An advocate for the Western Governor's University (WGU), Mr. Sklamberg also serves on WGU-Nevada's Advisory Board. Other long-standing community board positions include the following health and human service organizations and business groups: American Cancer Society's CEOs Against Cancer, the Council for a Better Nevada, Keep Our Doctors in Nevada (KODIN), Las Vegas Heals, Nevada Organ Donor Network (Chair-elect), Ronald McDonald House Charities of Greater Las Vegas (past-Chair), and the Board of Trustees with March of Dimes Nevada Chapter. Mr. Sklamberg holds a Master of Business Administration degree from the Olin School of Business at Washington University in St. Louis, Missouri, and a Bachelor of Arts degree from Hofstra University in New York.

Ronald M. Stewart, M.D., FACS, completed medical school and surgical residency at the University of Texas Health Science Center at San Antonio. He then completed a Trauma and Surgical Critical Care Fellowship at the University of Tennessee in Memphis. Returning to San Antonio in 1993, he established and built University Health System's trauma program. He currently serves as the Chair of the Department of Surgery, holding the Dr. Witten B. Russ Endowed Chair in Surgery at University of Texas Health San Antonio. Over the past three decades, he has actively led the development of an integrated civilian–military trauma system that serves all of South Texas, covering more than 26,000 square miles. In 2001, Dr. Stewart was appointed by then-Governor George W. Bush to the Texas Governors EMS and Trauma Advisory Council where he served for 15 years as the Chair of the Trauma Systems Committee. He was a founding member and the first Chair of the National Trauma Institute. For the past 20 years he has served on the American College of Surgeons (ACS) Committee on Trauma (COT), on which he recently concluded a term as chair. In 2013 Dr. Stewart was the recipient of the National Safety Council's Surgeons Award for Service to Safety, and the American College of Surgeon's Arthur Ellenberger Award for Excellence in State Advocacy. He is the immediate past President of the Southwestern Surgical Congress and the President of the Texas Surgical Society. During his tenure as the ACS COT Chair, he and the COT have spearheaded a plan to implement a National Trauma Action Plan aimed at eliminating preventable trauma deaths by (1) improving trauma systems, (2) increasing high quality trauma research, (3) increasing the quality of trauma patient data, and (4) advancing trauma education and training in both military and civilian settings. Additionally, he and the COT have worked to lead an approach to firearm injury prevention that has encouraged a collegial, professional and substantive dialogue from surgeons and citizens from all points of view with the goal of reducing the burden of firearm injury and death.

William Craig Vanderwagen, M.D., RADM (USPHS), is a family physician who retired as a Rear Admiral in the U.S. Public Health Service (USPHS) in 2009. He served for 25 years in the Indian Health Service, the federal program of medical and public health services for American Indians and Alaska Natives. During this period he also served as the lead health official at a number of disasters including: medical care for Kosovar refugees (1999); advisor to the Afghan Ministry of Health (2002); director of public health and advisor to the Iraq Ministry of Health (2003–2004); the U.S. Navy Ship Mercy response to the 2004 tsunami; and commander of the public health and medical response to Hurricanes Katrina/Rita. Dr. Vanderwagen's last federal assignment (2006–2009) was as the Assistant Secretary for Preparedness and Response at the Department of Health and Human Services

(HHS). He was responsible for leading all federal public health and medical assets in disaster response and, responsibility for guiding the $11 billion HHS medical countermeasure advanced development program to address chemical, biological, radiological, and nuclear threats, which now has more than 100 products in the development pipeline. Dr. Vanderwagen is a Director and General Manager of East West Protection, a Potomac, Maryland-based firm specializing in public health and medical preparedness, detection, response, and command and control systems for CBRN threats and other disasters. He is also the Chairman of the Board at the International Center for Infectious Disease. He is vice chairman of the Board at the Vaccine and Infectious Disease Organization-Intervac, a Canadian vaccine research and development company. He is also a senior partner at Martin, Blanck, and Associates, a consulting firm of retired General and Flag officers specializing in military health matters. He is a frequent public speaker on biodefense, public health preparedness, and leadership.

Jody R. Wireman, Ph.D., M.S.P.H., M.P.A., CIH, DABT, has more than 25 years of experience as a public health professional, manager, and educator. He is currently the Director of Force Health Protection (FHP) at the North American Aerospace Defense Command (NORAD) and U.S. Northern Command (USNORTHCOM). Dr. Wireman provides leadership, management, and expertise in occupational, environmental, and chemical, biological, radiological, and nuclear (CBRN) FHP and Health Service Support to the Commander and Surgeon. He also supports other agencies for refinement of medical and public health needs for homeland defense and civil support missions. Dr. Wireman and his staff directly support development of deliberate and crisis action plans for NORAD and USNORTHCOM for mission assurance and consequence management responses throughout North America. Dr. Wireman has played a key role in developing the Department of Defense's (DoD's) medical capabilities and associated Commander's estimate for homeland defense and civil support missions. This includes identifying FHP requirements and guidance for the 18,000-member DoD Chemical, Biological, Radionalical, and Nuclear Response Enterprise and follow on forces. Dr. Wireman contributes to his profession through participation in public health organizations, authoring journal and book articles, and volunteering as an expert on human and environmental health projects. Prior to his current assignment, Dr. Wireman was a Deputy Division and Branch Chief at NORAD and USNORTHCOM, Deputy Branch Chief, toxicologist, and human and ecological health scientist at the U.S. Air Force School of Aerospace Medicine and served as a consultant. His previous efforts focused on worker health protection and environmental restoration of radiologically- and chemically-contaminated hazardous waste sites. Dr. Wireman worked at various levels of the government from local to National levels (including Tribal Nations and regulators), and local community members and academia in identifying risks and cleanup options, and evaluating toxicity of chemicals and products of military significance. Dr. Wireman has degrees in Toxicology (Texas Tech University), Environmental Health (University of Alabama at Birmingham), Public Management (Harvard University), and Occupational Safety (Indiana University of Pennsylvania). He is also is a Diplomat of the American Board of Toxicology and is a Certified Industrial Hygienist.

David Witt, M.D., FIDSA, CIC, is an Infectious Disease Specialist with Kaiser Permanente in Northern California. He attended medical school at the University of Michigan and residency and fellowship at the Boston City Hospital in Boston, Massachusetts. He has been board certified in Internal Medicine, Infectious Diseases, Emergency Medicine, and certified in Tropical Medicine and Infection Control. He is on the faculty of the Schools of Medicine and Nursing at the University of California, San Francisco. Dr. Witt chairs the Clinical Workgroup for National Healthcare Continuity, which addresses Emergency Preparedness for Kaiser Permanente. He works with the MENTOR Initiative, a British nongovernmental organization dedicated to reducing the burden of vector-borne diseases in humanitarian crises. As Clinical Director for the MENTOR Initiative, he has responded to multiple disasters, including the Indonesian Tsunami, the Tana River floods in Kenya and Somalia, Cyclone Nargis in Myanmar and the earthquake in Haiti. He was Clinical Director for the second team of Kaiser Permanente's response to Hurricane Katrina. Previously, he was Medical Director of South San Francisco Fire Department and served as a Technical Auditor for the U.S. State Department, working for the International Science and Technology Center, an activity of seven countries and the European Union dedicated to non-proliferation of the former Soviet Union's nuclear, chemical, and biologic weapons and their technology.

Joseph L. Wright, M.D., M.P.H., is the recently appointed Professor and Chairman of Pediatrics at the Howard University College of Medicine in Washington, DC. He previously served as Senior Vice President within the Children's National Health System, where he provided strategic leadership for the organization's advocacy mission, public policy positions, community partnership initiatives, and served as founding director of the Child Health Advocacy Institute. He maintains adjunct appointments as professor of emergency medicine and health policy and management at the George Washington University Schools of Medicine and Public Health, as well as professor of family science at the University of Maryland School of Public Health. He served 17 years as the inaugural state pediatric medical director within the Maryland Institute for Emergency Medical Services Systems, in addition to 8 years as principal investigator and medical director of the federally funded Emergency Medical Services for Children (EMSC) National Resource Center. Academically, Dr. Wright is among the original cohort of board-certified pediatric emergency physicians with scholarly interests that include injury prevention, prehospital pediatrics, and the needs of underserved communities. He has contributed more than 100 publications to the scientific literature, been invited to serve two dozen visiting professorships, and is currently principal investigator of the National Institutes of Health-funded DC-Baltimore Research Center on Child Health Disparities. Dr. Wright is an elected member of both Alpha Omega Alpha and Delta Omega, the nation's highest medical and public health honor societies, respectively, the American Pediatric Society, and was recently inducted into the Academy of Medicine of Washington. Dr. Wright is current chair of the American Academy of Pediatrics (AAP) Committee on Pediatric Emergency Medicine after having served as inaugural chair of the AAP Violence Prevention Subcommittee. He is also currently chairing the AAP Task Force on Addressing Bias and Discrimination. Dr. Wright has been recognized throughout his career for his advocacy work highlighted by receipt of two lifetime achievement awards from the AAP for distinguished contributions to the disciplines of injury prevention and emergency medicine, respectively. He is also a recipient of the prestigious Shining Star Award from the Los Angeles–based Starlight Foundation for outstanding service to communities of color. Dr. Wright provides national leadership through advisory and governance service to several organized medicine bodies including the Association of American Medical Colleges, the American Hospital Association, the March of Dimes, the National Academy of Medicine, the National Highway Traffic Safety, and a recently completed term as an Obama administration appointee to the Pediatric Advisory Committee of the Food and Drug Administration. He regularly presents invited expert testimony before Congress, state and municipal legislative bodies, has made numerous media appearances, and lectures widely to both professional and lay audiences. Dr. Wright earned a B.A. from Wesleyan University, his M.D. from Rutgers New Jersey Medical School, and a Master's of Public Health in Administrative Medicine & Management from George Washington University.

Kevin Yeskey, M.D., currently serves as a senior advisor to the Assistant Secretary for Preparedness and Response (ASPR) in the Department of Health and Human Services (HHS). He previously served in ASPR as Deputy Assistant Secretary for Preparedness and Response. Most recently, Dr. Yeskey also served as a Senior Advisor for Emergency Public Health with MDB, Inc. He is former director of the Center for Disaster and Humanitarian Assistance Medicine in the School of Medicine at the Uniformed Services University of the Health Sciences (USUHS). A graduate of Brown

University, Dr. Yeskey received his medical degree from USUHS, and is board-certified in emergency medicine. While on active duty with the U.S. Public Health Service (USPHS) he served in a variety of disaster response positions, including duty as senior medical policy advisor in the Response Division of the Federal Emergency Management Agency, as director of the HHS Office of Emergency Response, and as director of the Bioterrorism Preparedness and Response Program of the Centers for Disease Control and Prevention. He also appeared before Congress as an expert witness in the field of bioterrorism preparedness. The author of several textbook chapters and journal articles, Dr. Yeskey is the recipient of the Surgeon General's Medal for Exemplary Service and of several USPHS and Department of Defense Awards.

Richard Zuschlag, a native of Greensville, Pennsylvania, graduated from the Capitol Institute of Technology in Washington, DC, in 1970. In 1971, he joined with two friends to form Acadian Ambulance Service in Lafayette, Louisiana with two ambulances and eight Vietnam veteran medics. Under Mr. Zuschlag's leadership, Acadian Ambulance has become the nation's largest employee-owned ambulance service, with more than 4,300 employees, 550 ambulances, 8 helicopters, and 4 airplanes serving a population of more than 20 million in 34 parishes in Louisiana, 37 counties in Texas and Jackson County, Mississippi. The company has also expanded to include Acadian Air Med, Acadian Total Security, Executive Air Charter, the National EMS Academy, and Safety Management Systems. Acadian Ambulance received national recognition for their incredible response to Hurricanes Katrina and Rita in 2005. When Hurricane Katrina destroyed telephone and cellular communication in Southeast Louisiana, Acadian had the only working communications network and worked closely with the State of Louisiana, the National Guard and the federal government in coordinating the massive rescue and response efforts in the New Orleans area. More recently, Acadian played a key role in response to the unprecedented flooding in Southeast Texas caused by Hurricane Harvey. In addition to mobilizing its vast resources, Acadian's medical teams were instrumental in assisting the State of Texas in setting up and operating temporary emergency rooms and clinics in several locations. Mr. Zuschlag credits much of Acadian's success to the public–private partnerships he developed. Working with governmental agencies on the local, state and federal level, Mr. Zuschlag's experience and expertise has led to many advancements in systems and infrastructure to greatly improve the response to mass casualty incidents and disasters. In recognition of his contributions to the emergency medicine field, Mr. Zuschlag was awarded the Lifetime Achievement Award at the 2016 Pinnacle EMS Leadership Conference.